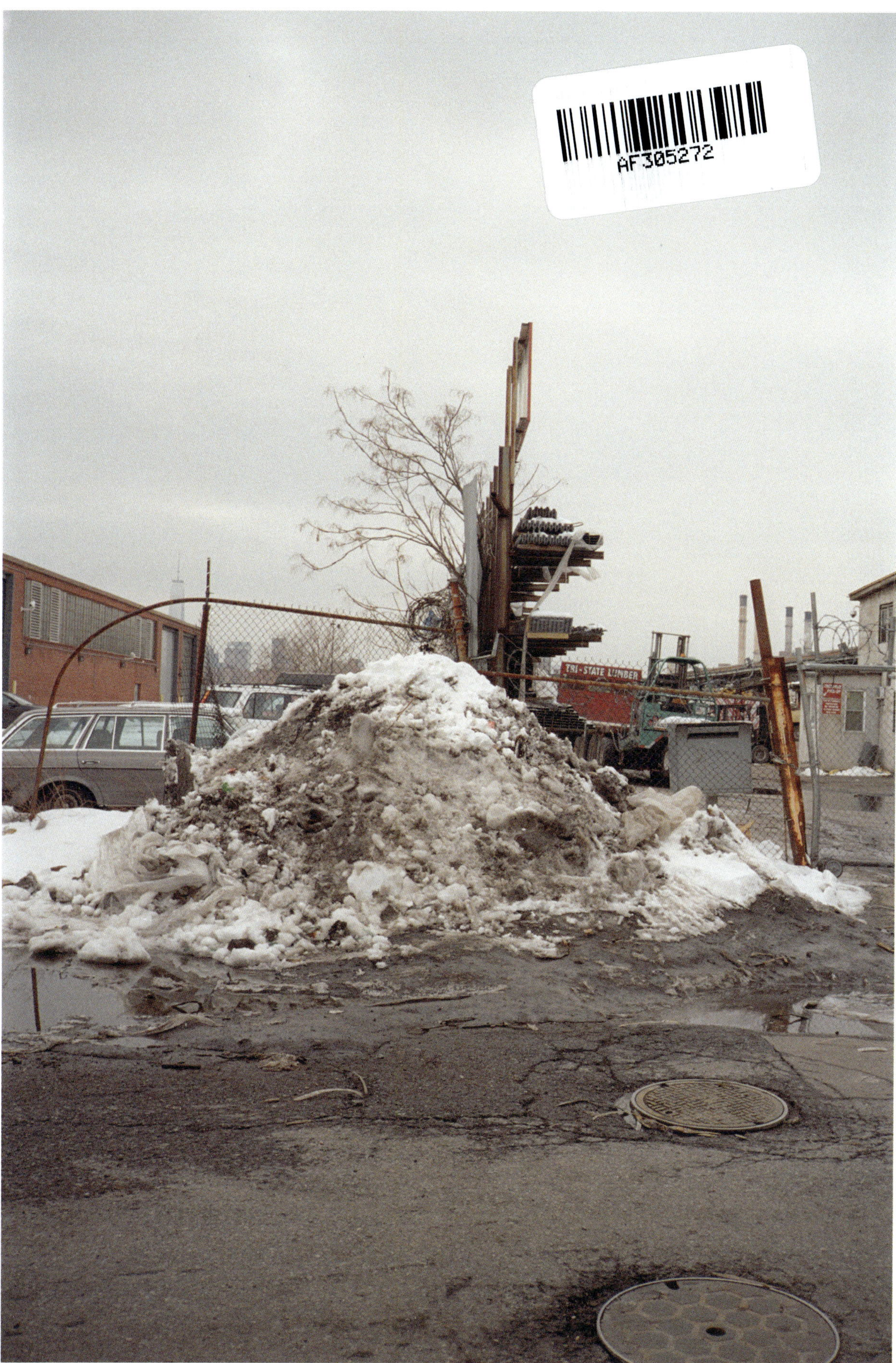
AF305272
TRI-STATE LUMBER

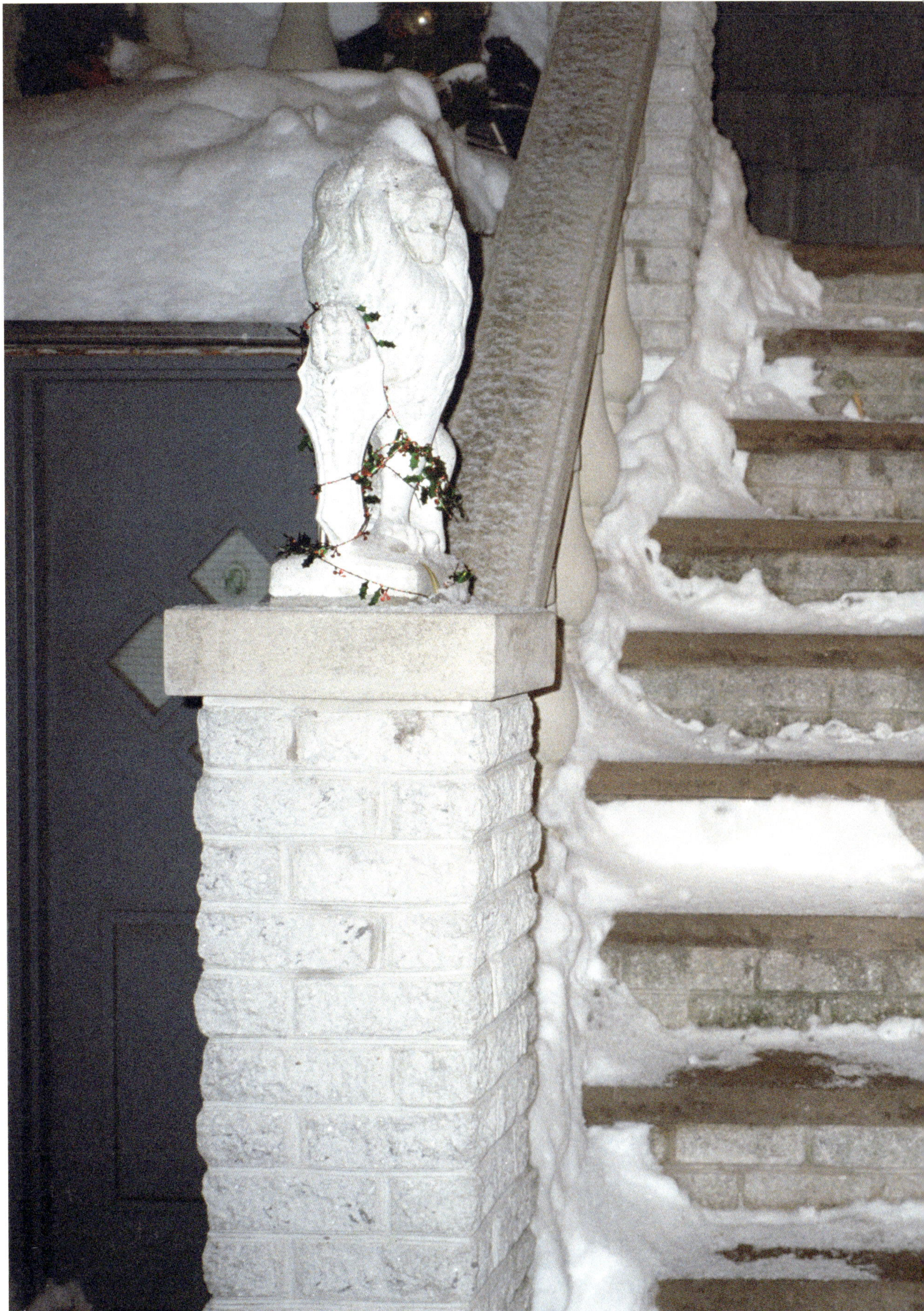

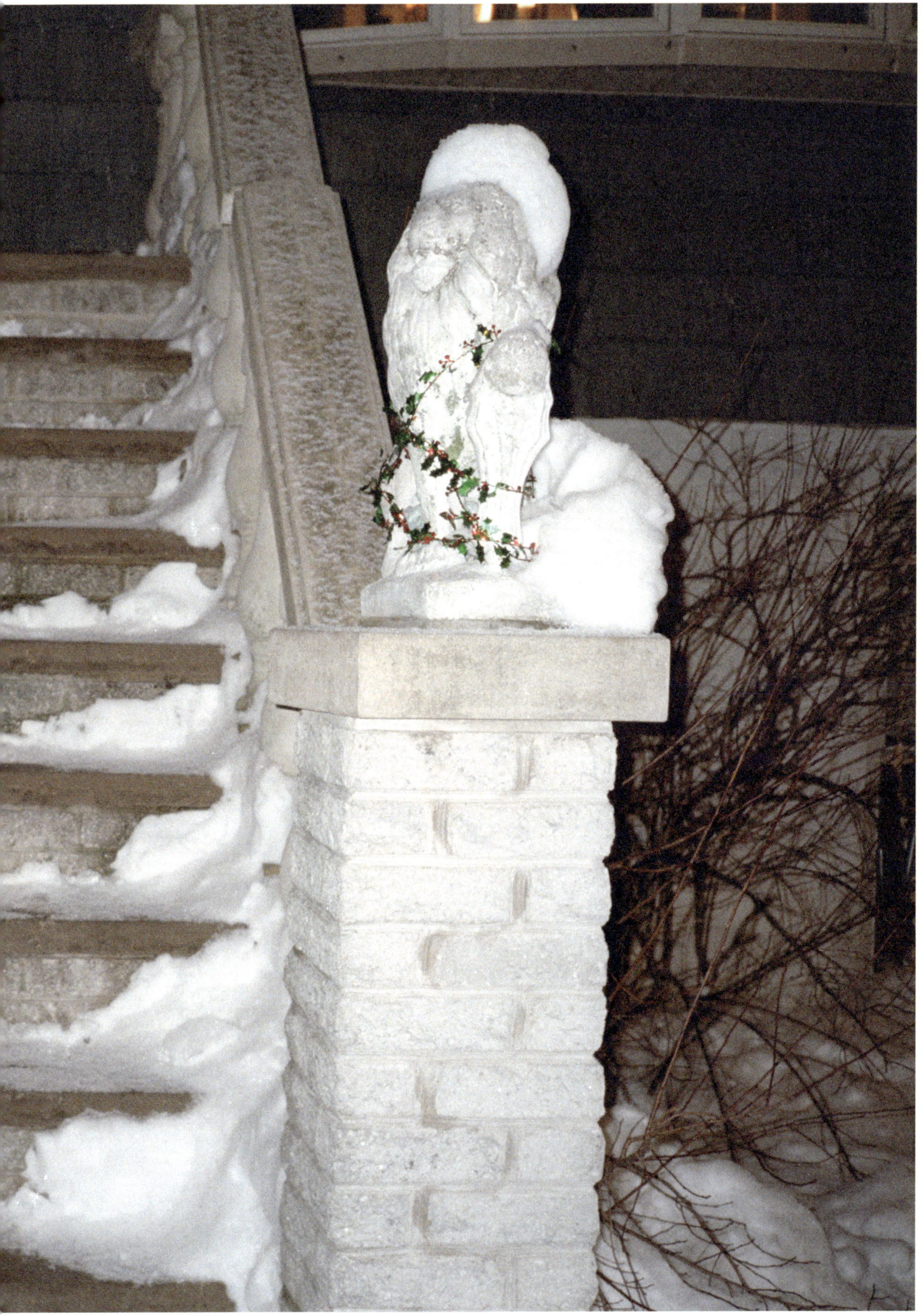

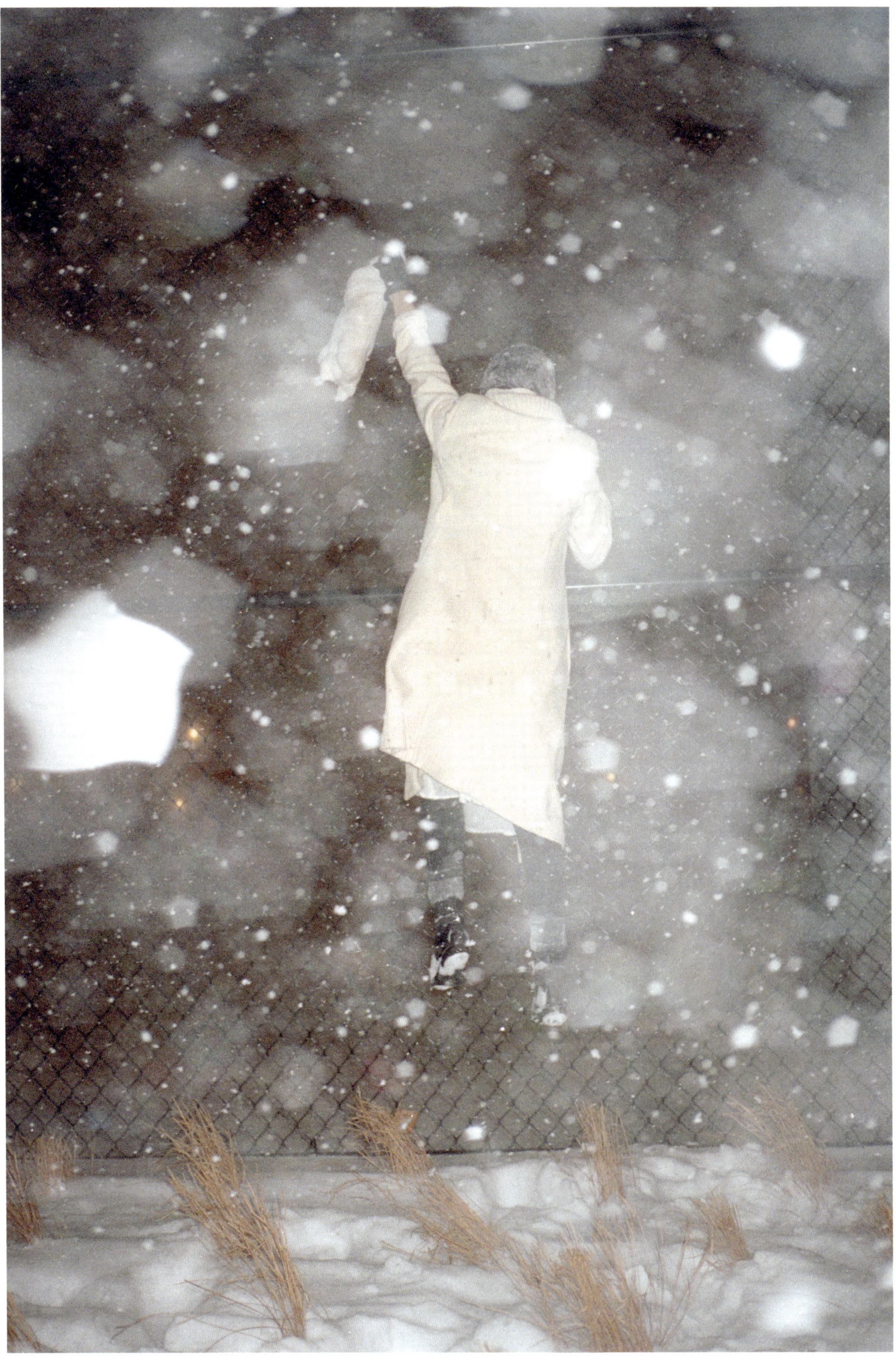

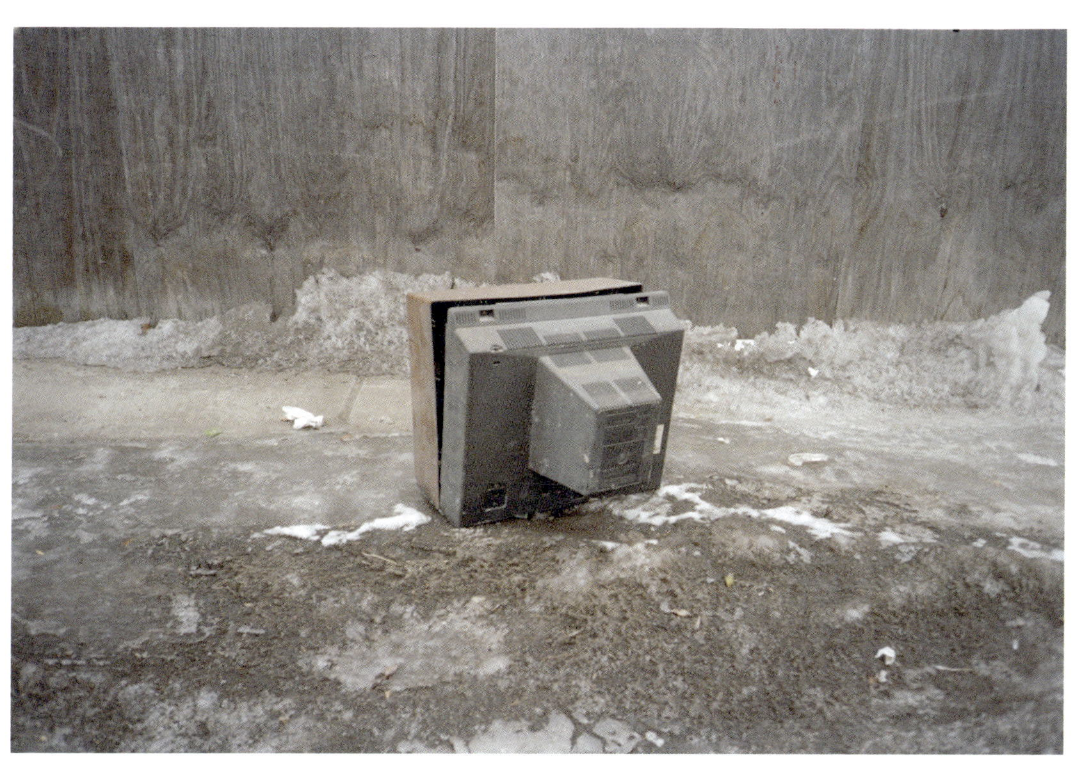

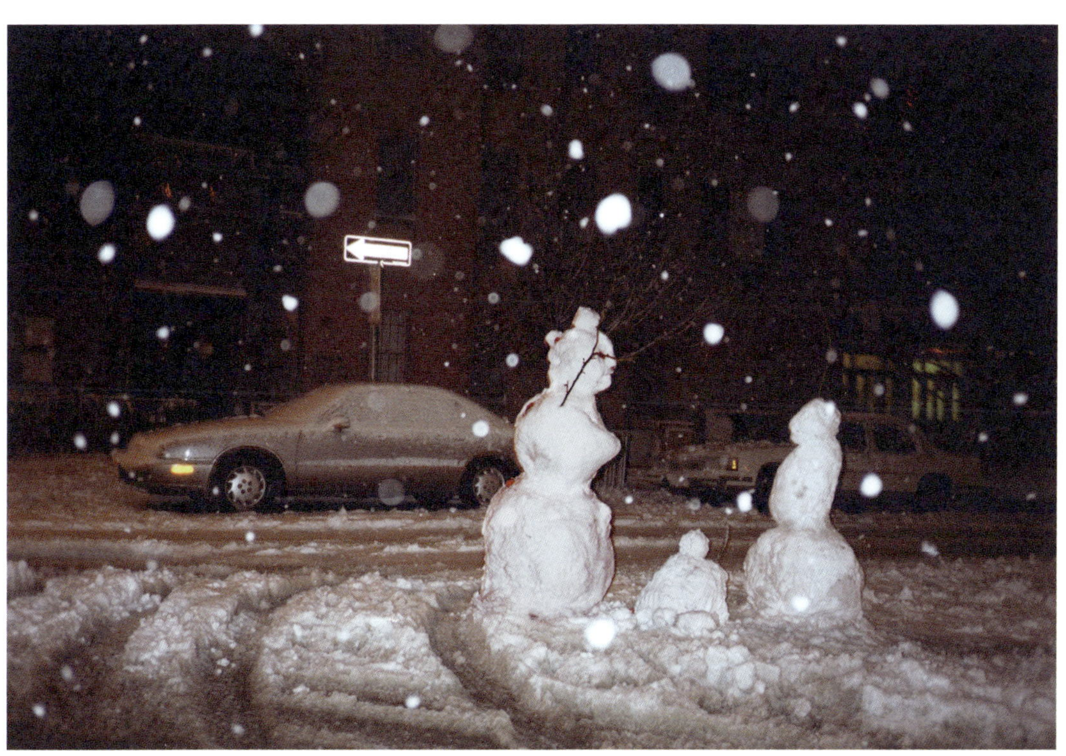

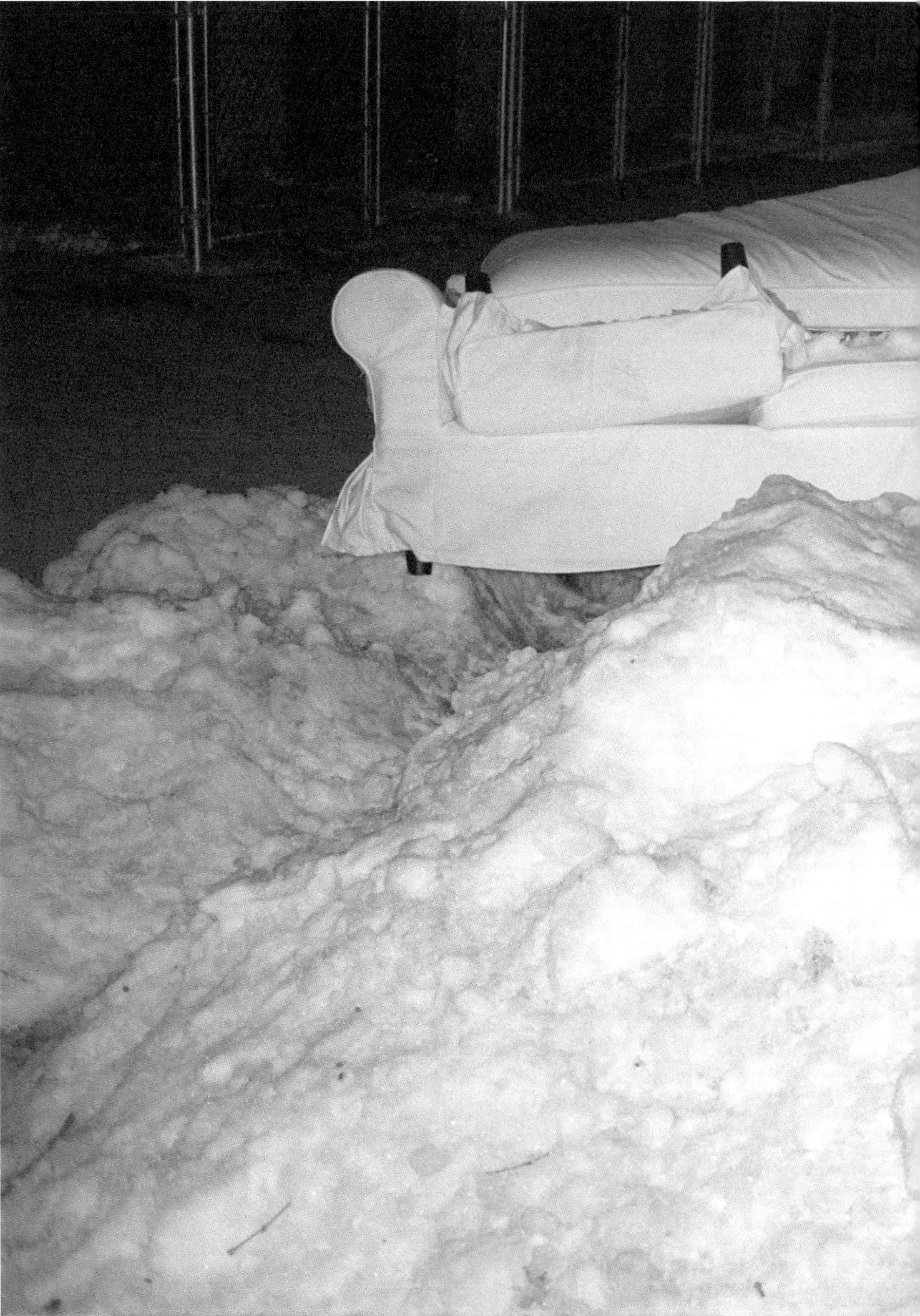

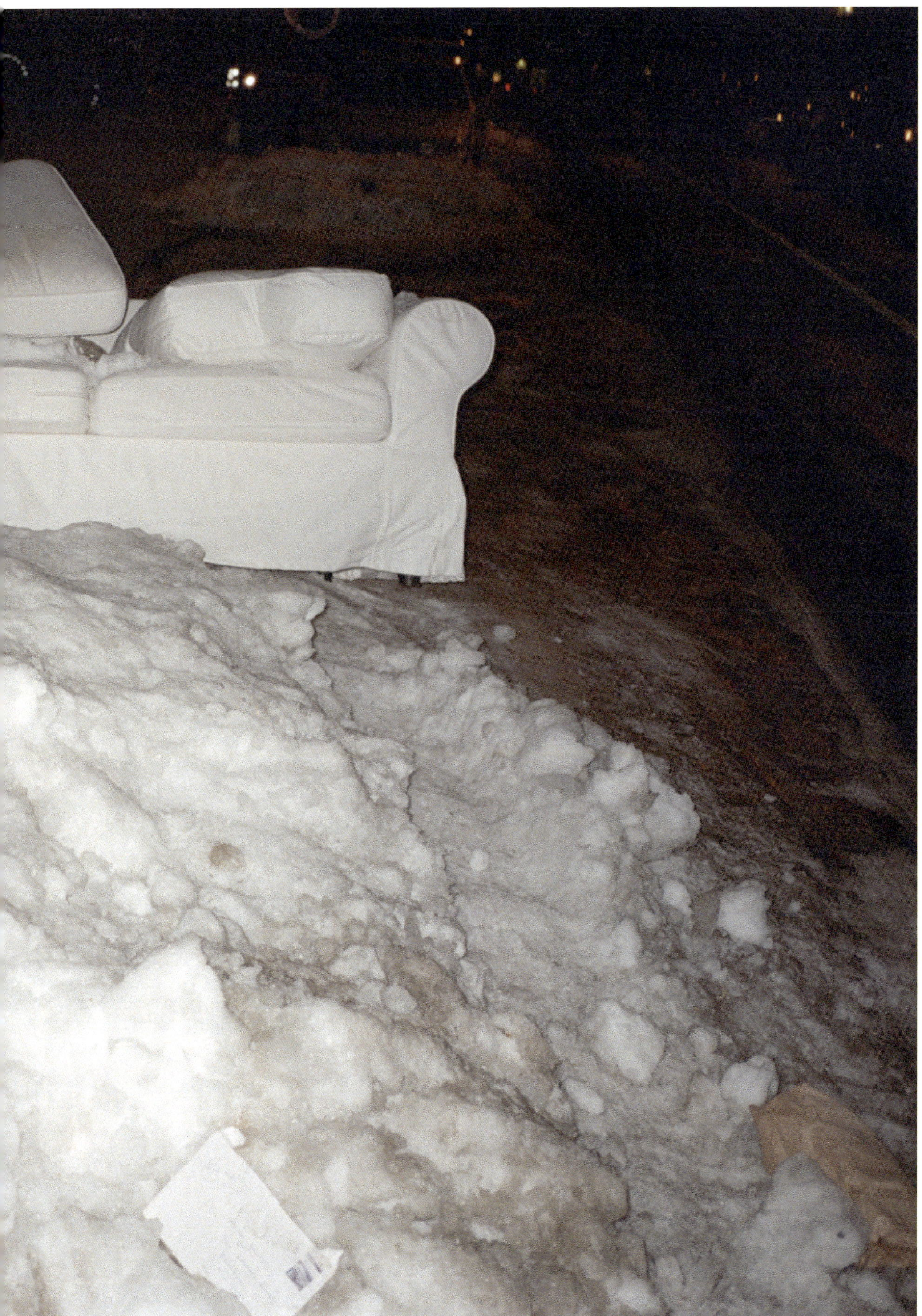

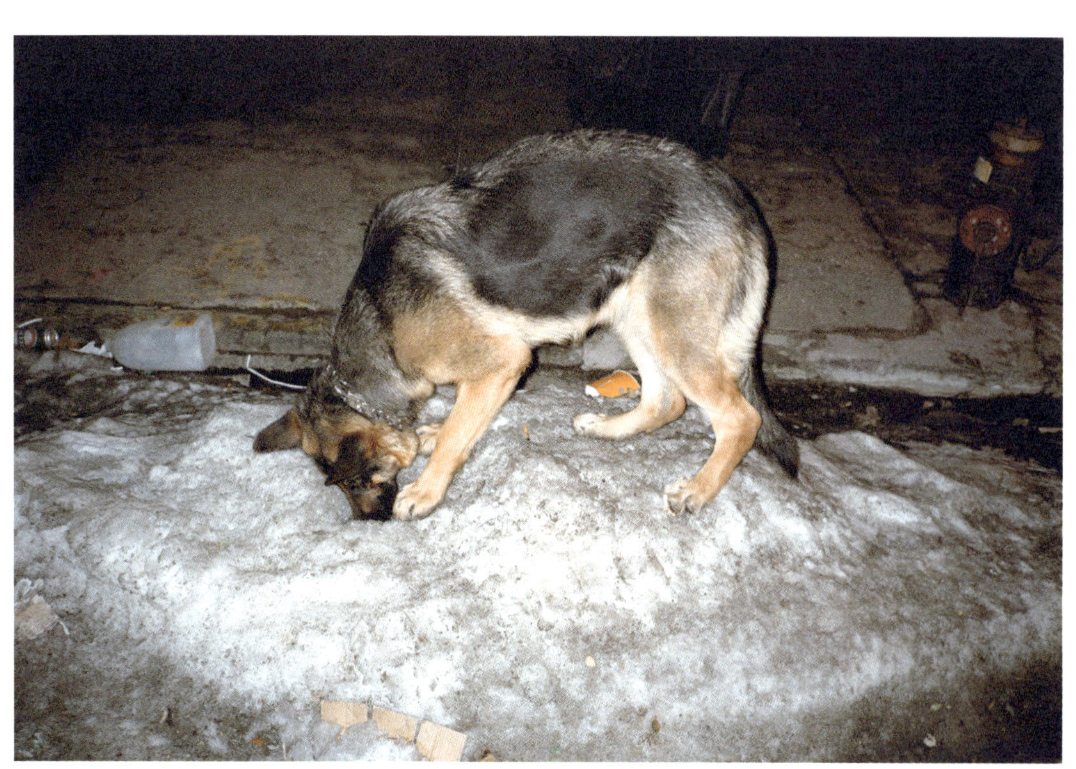

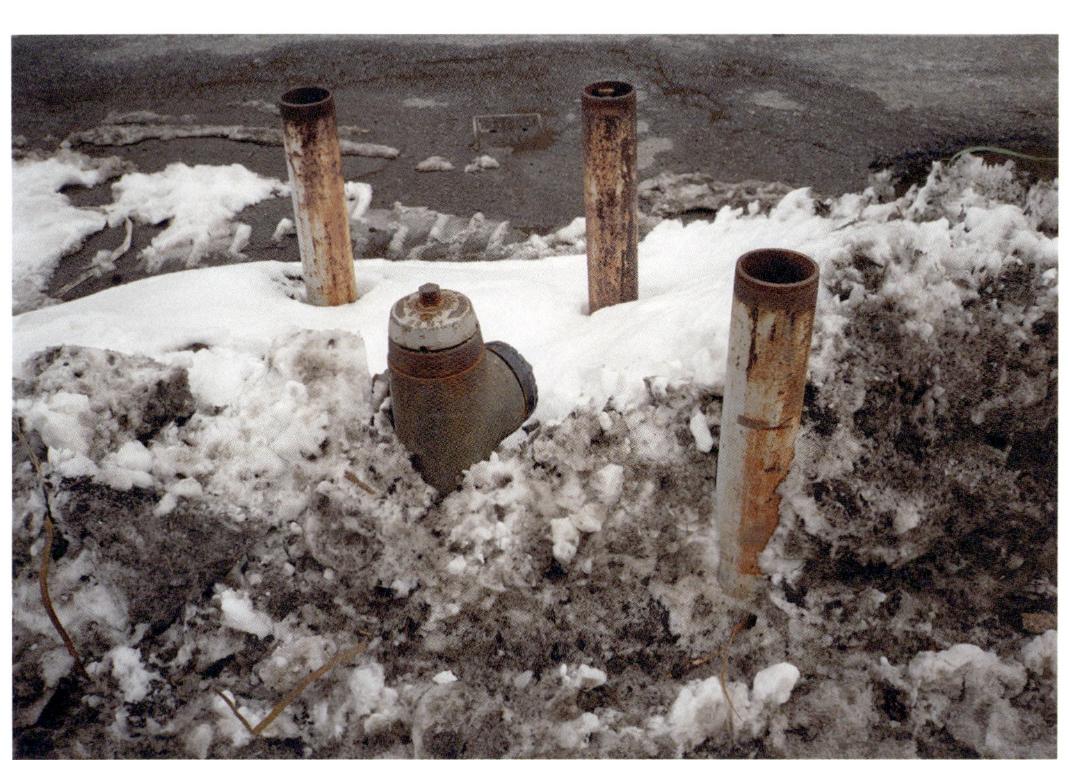

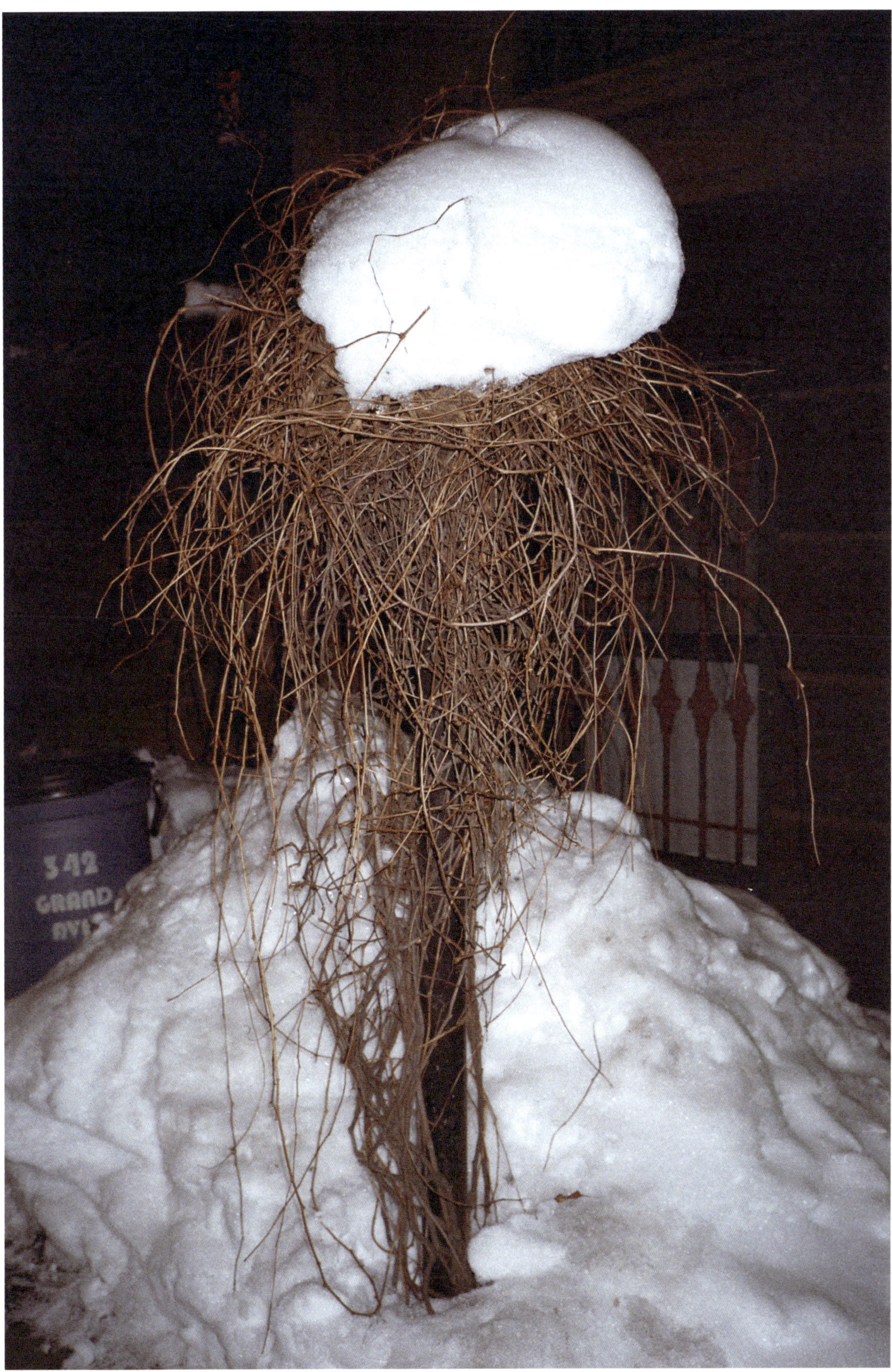

342
GRAND

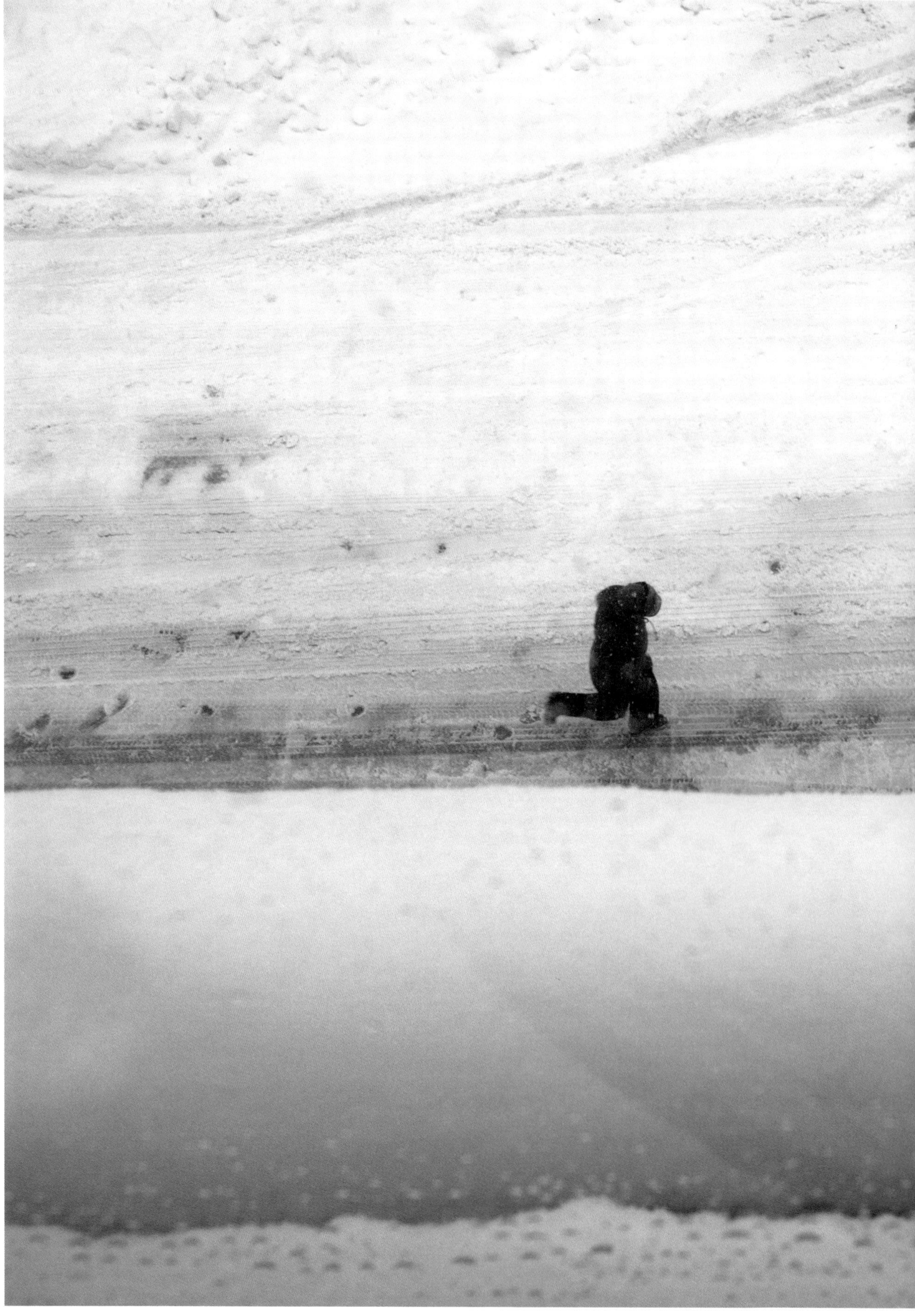

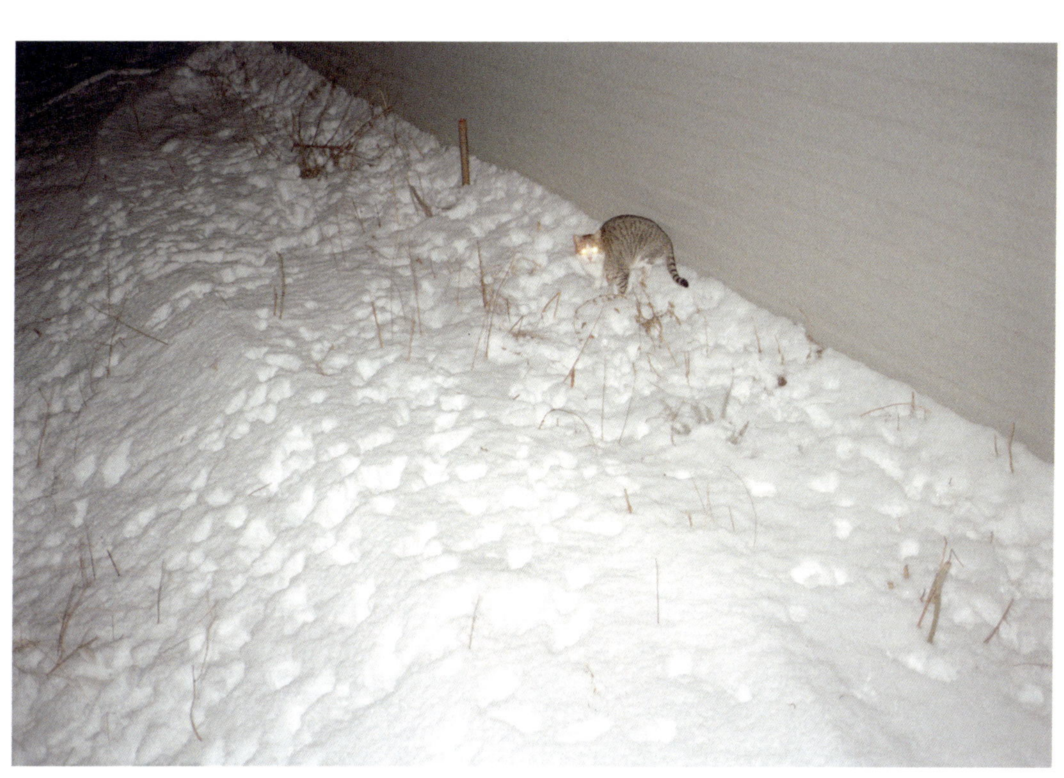

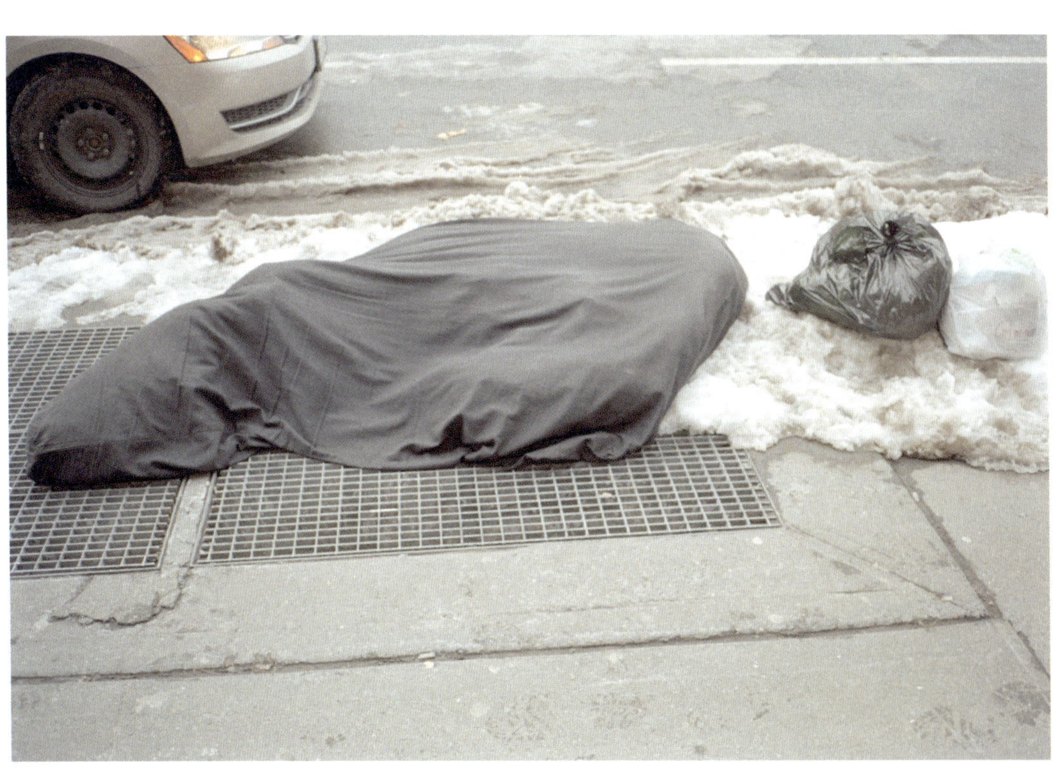

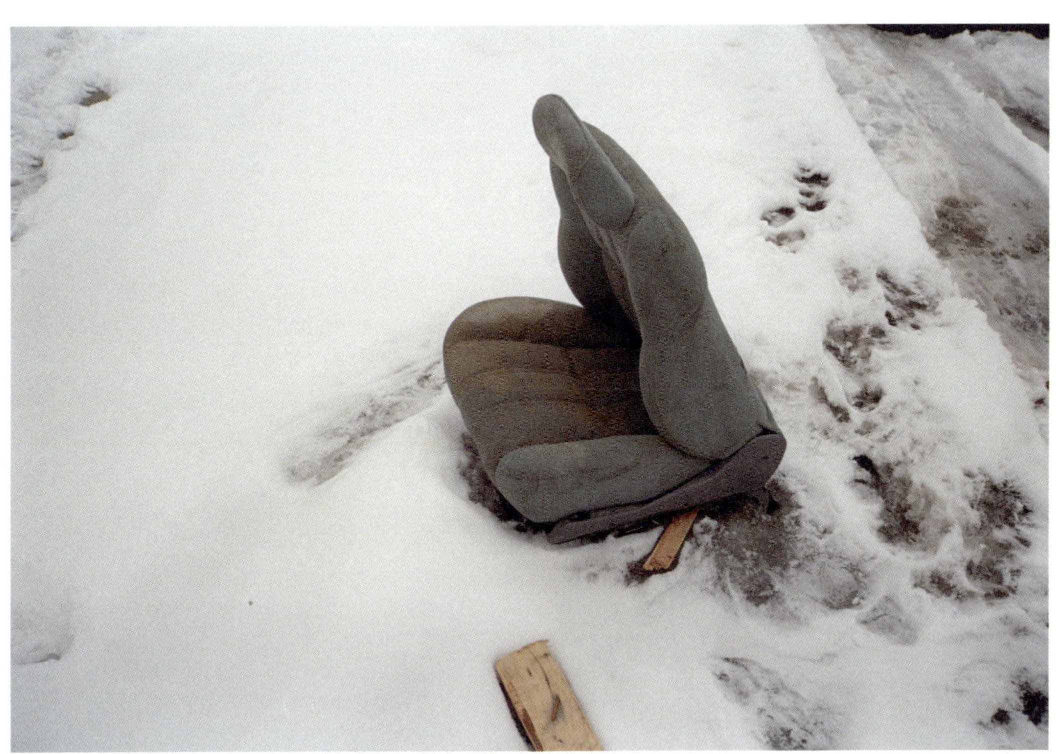

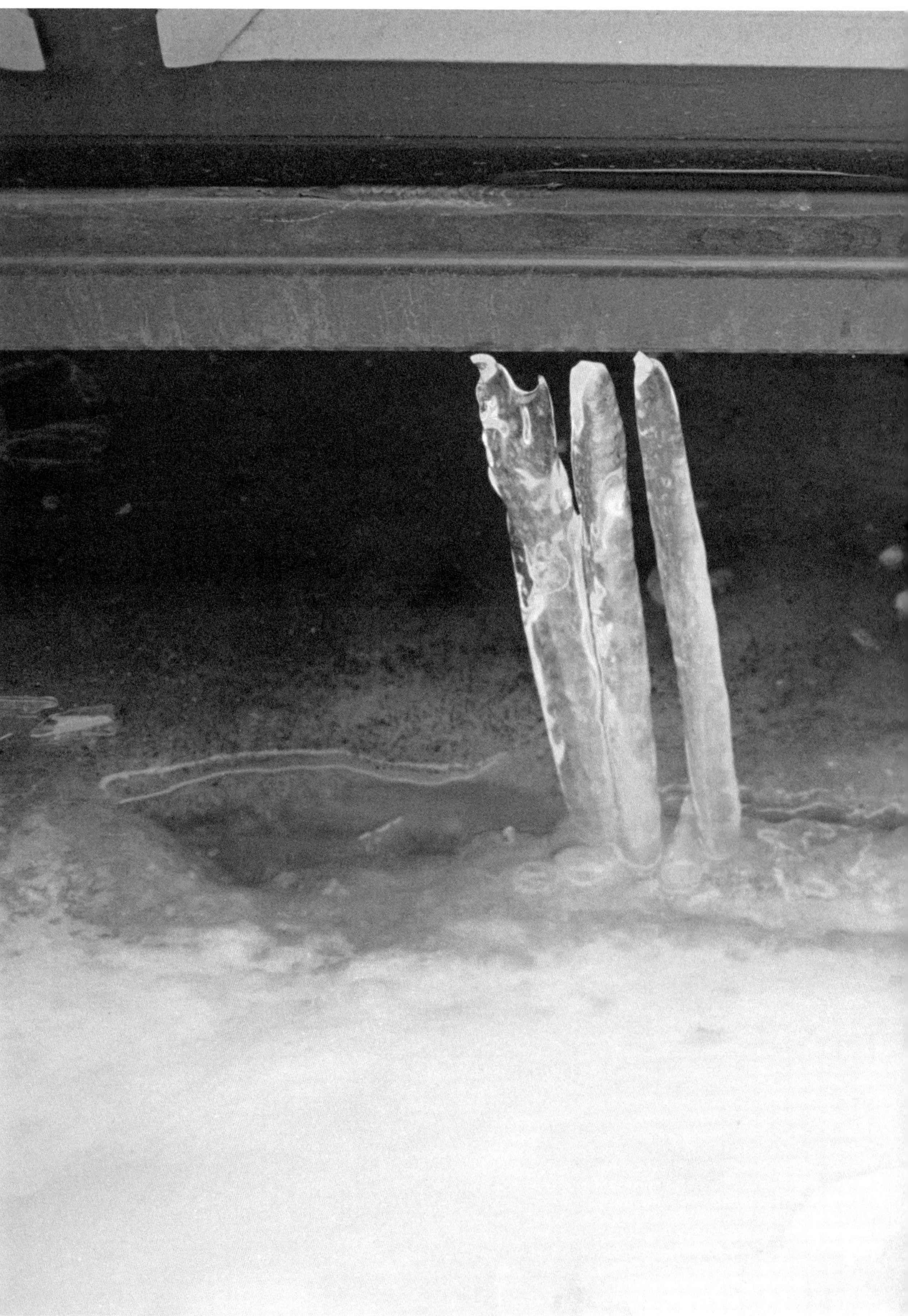

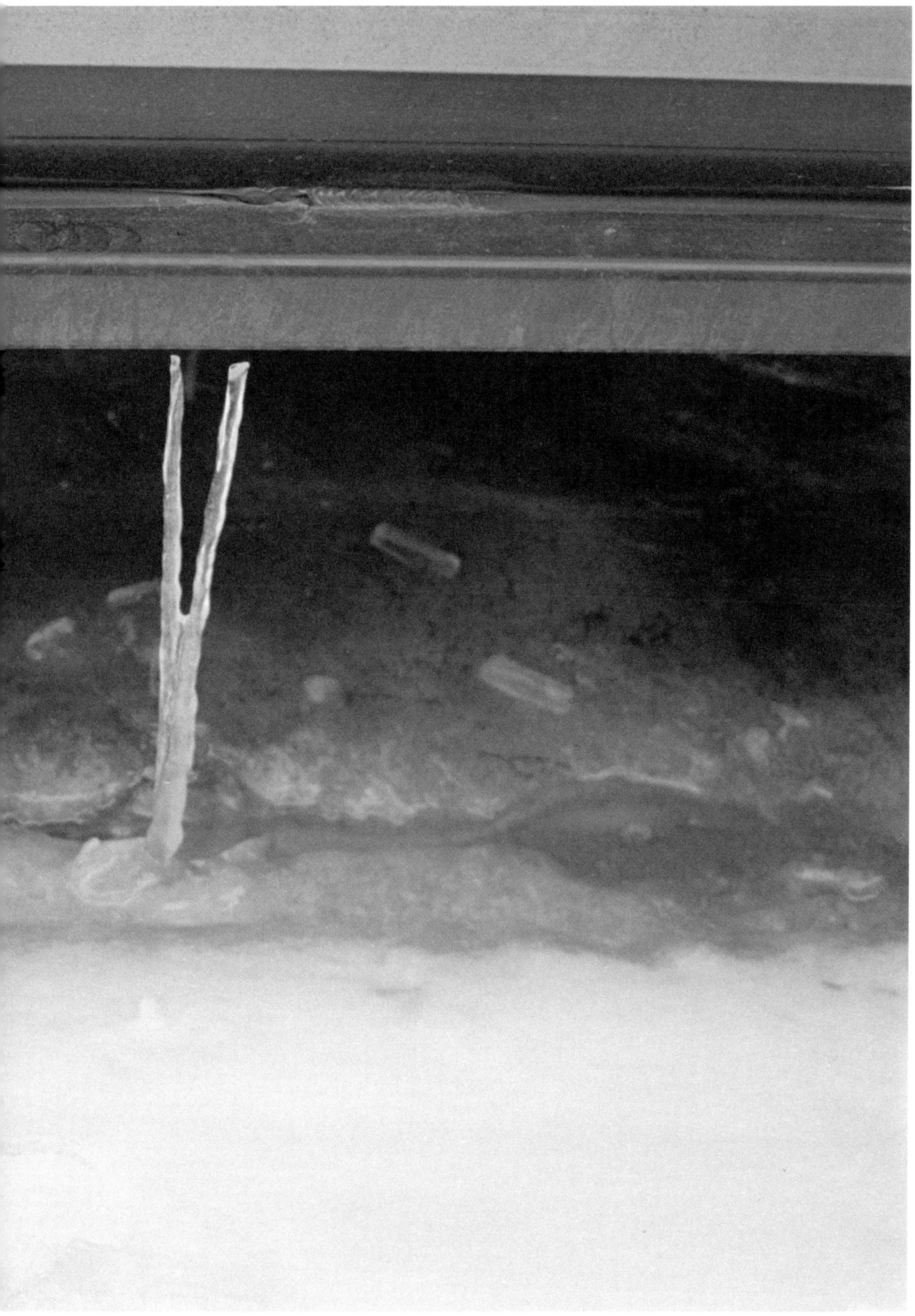

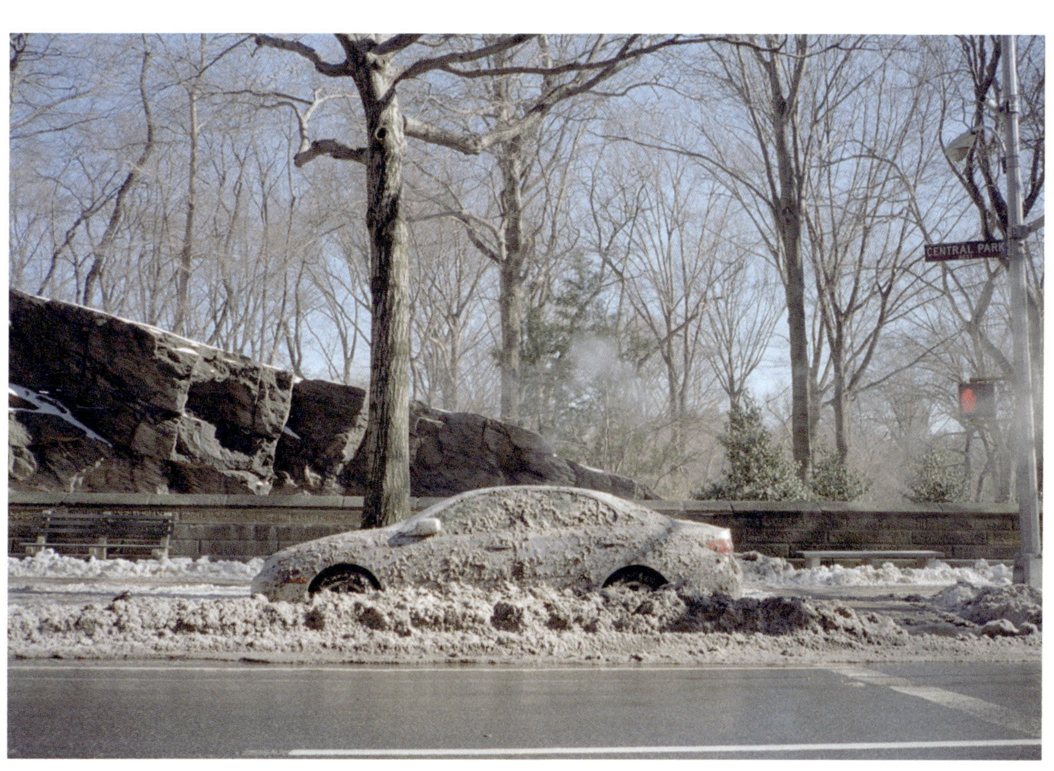
CENTRAL PARK

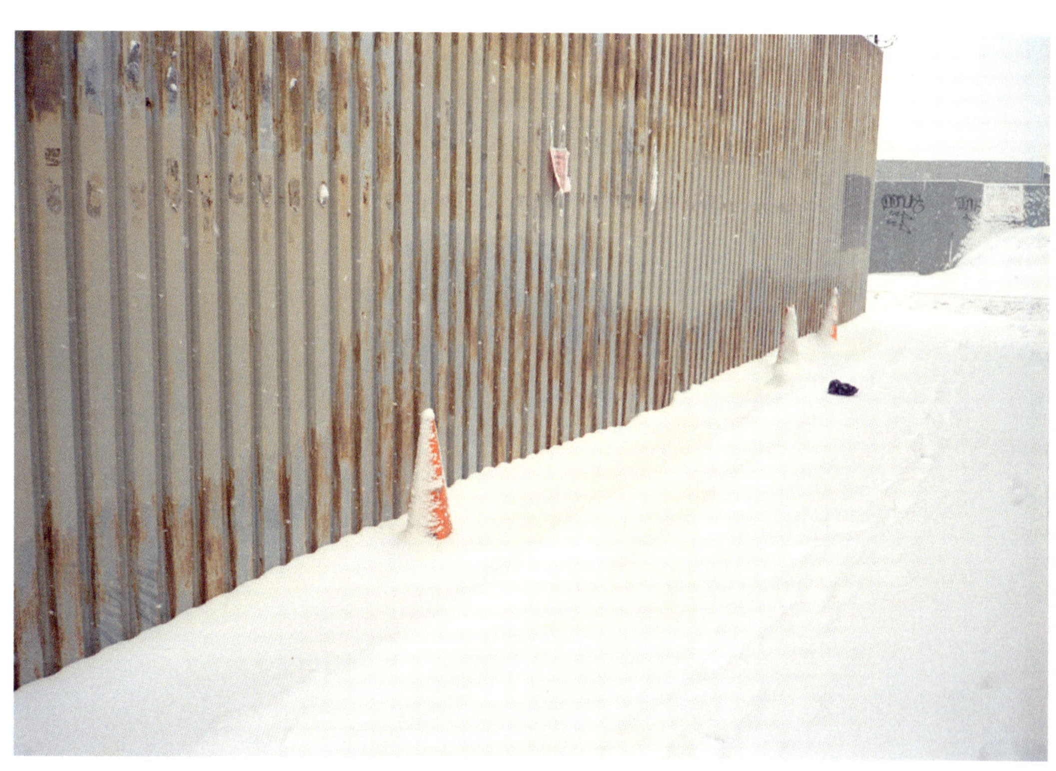

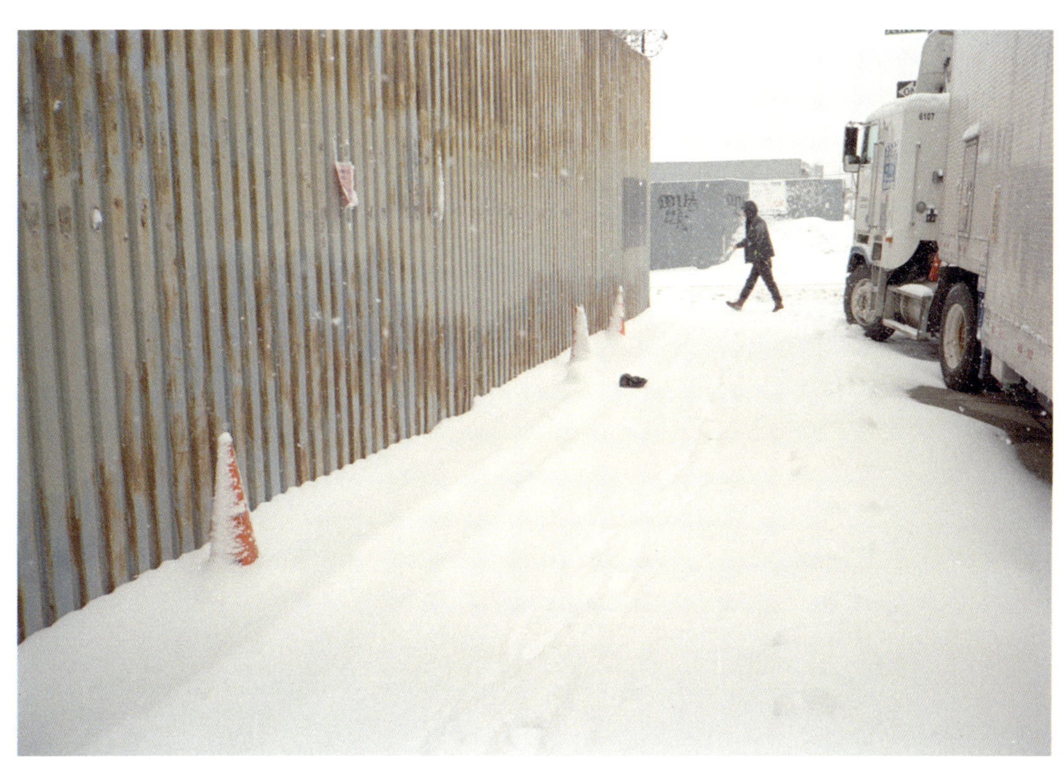

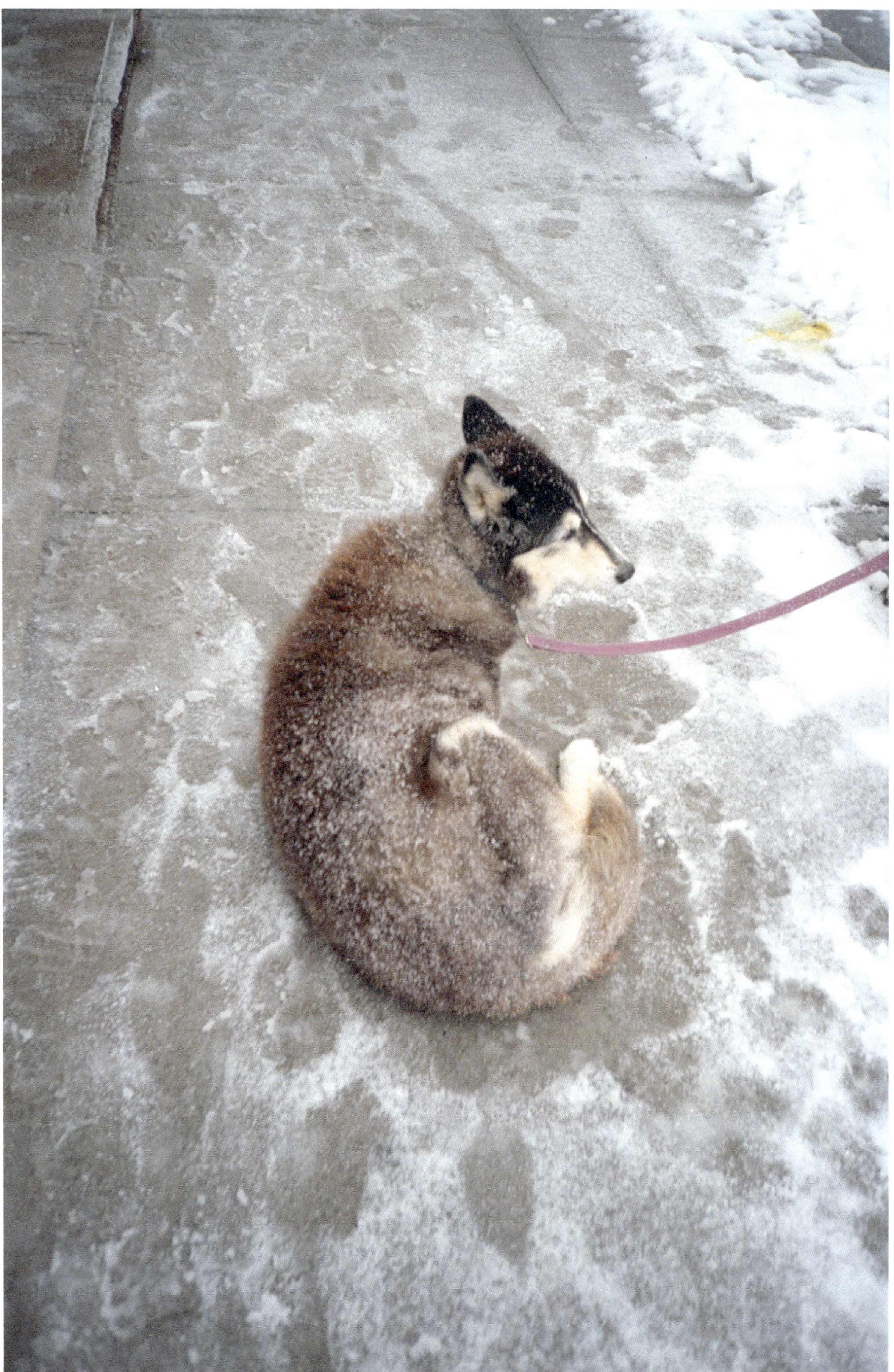

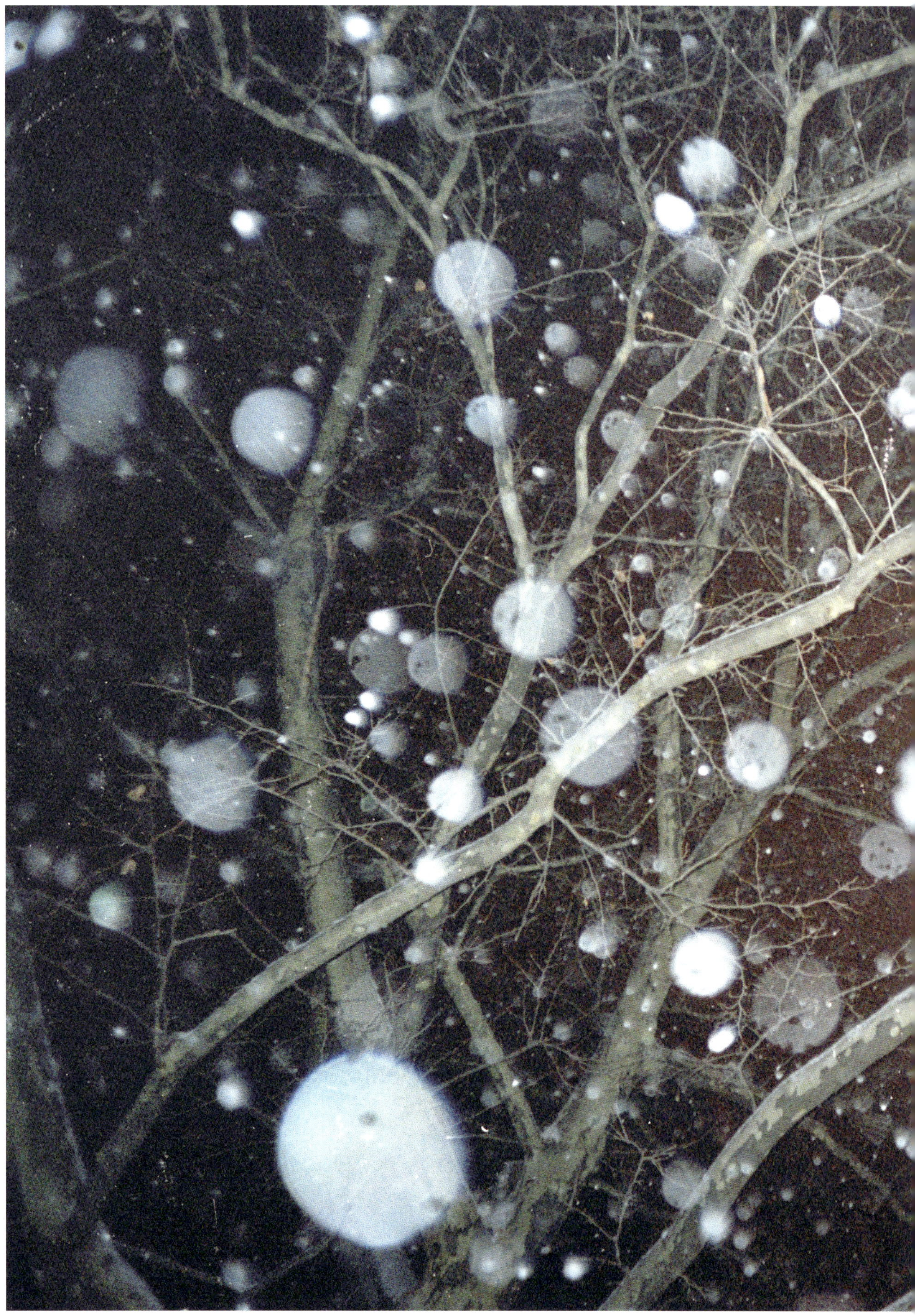

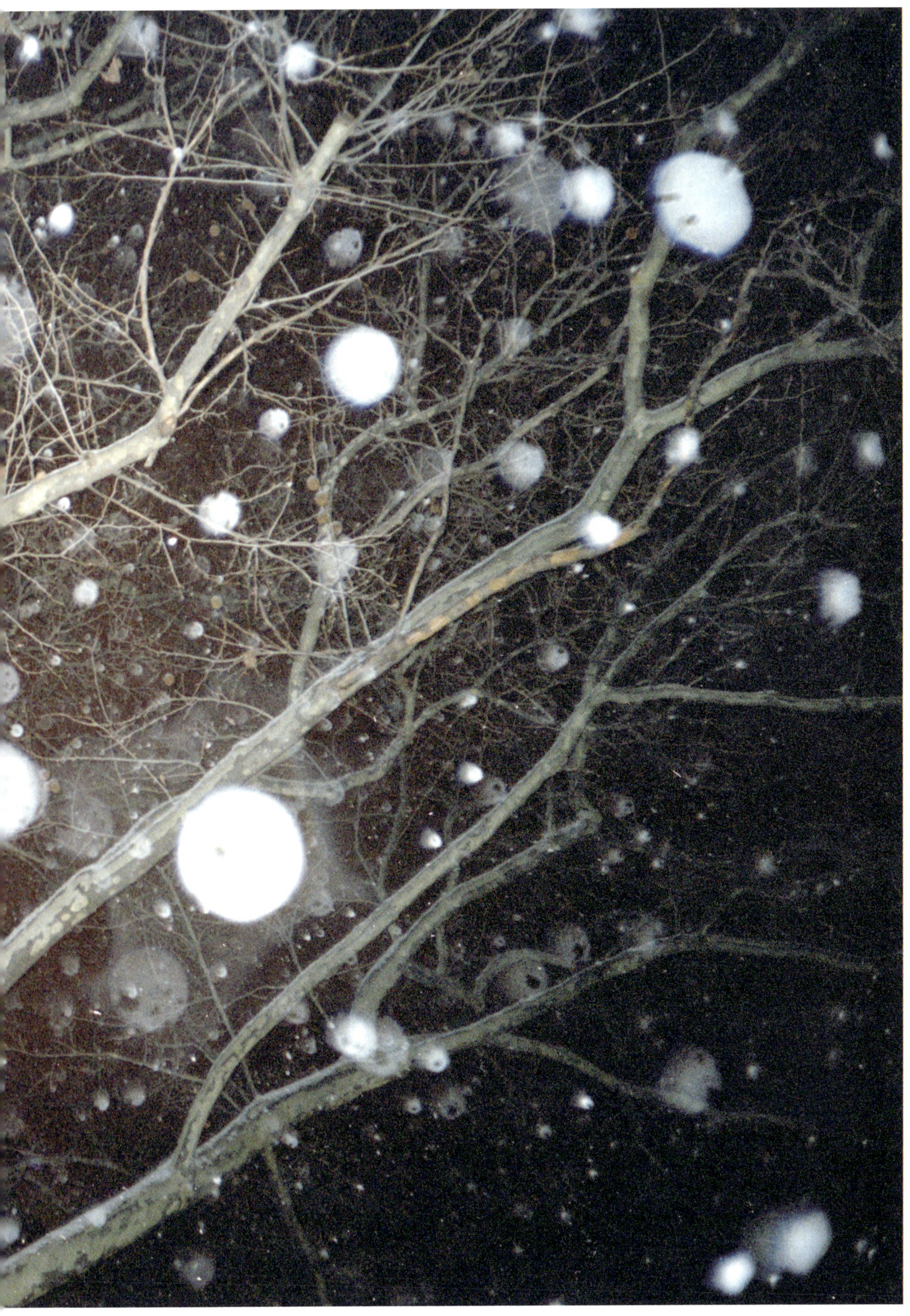

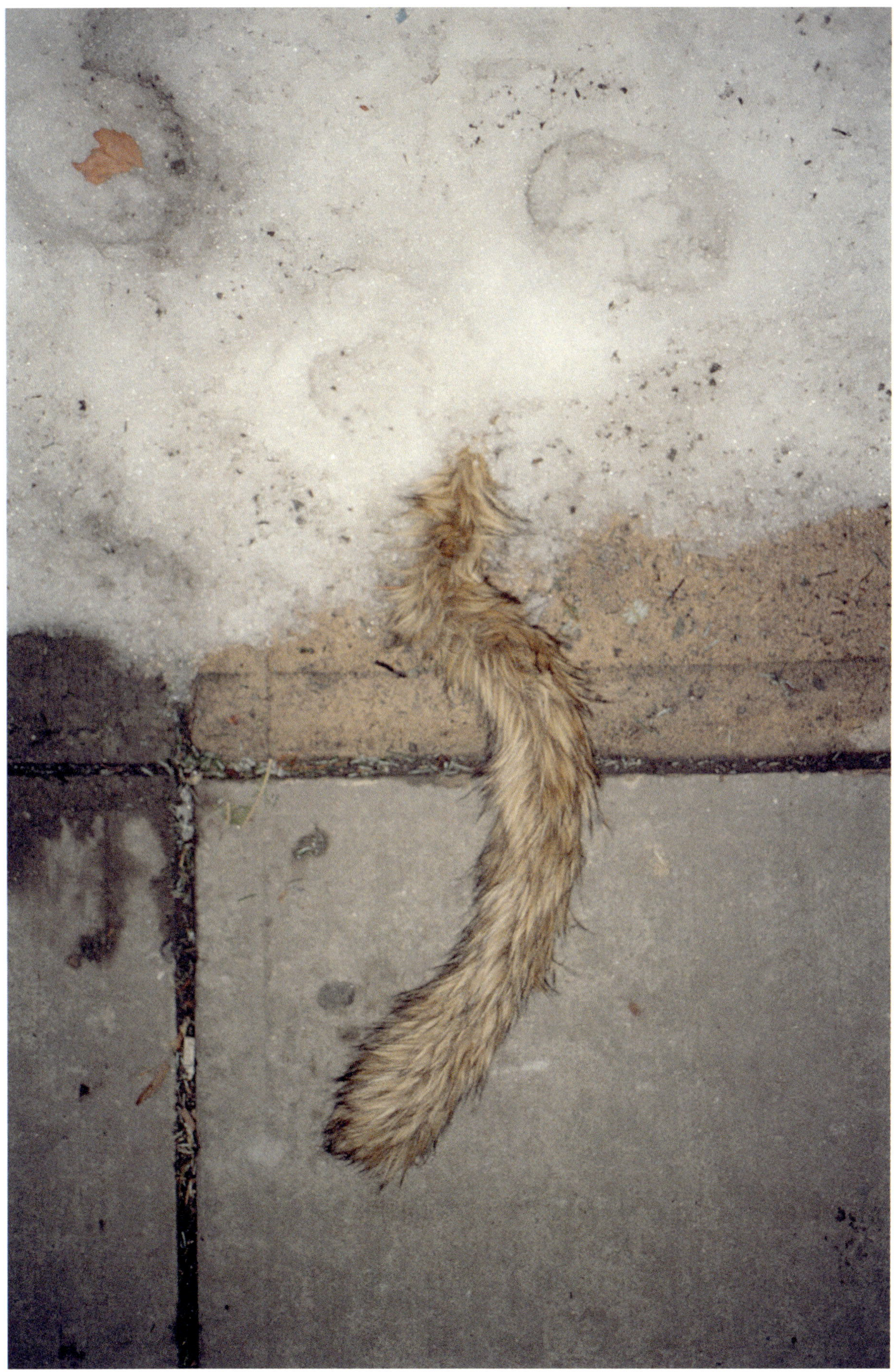

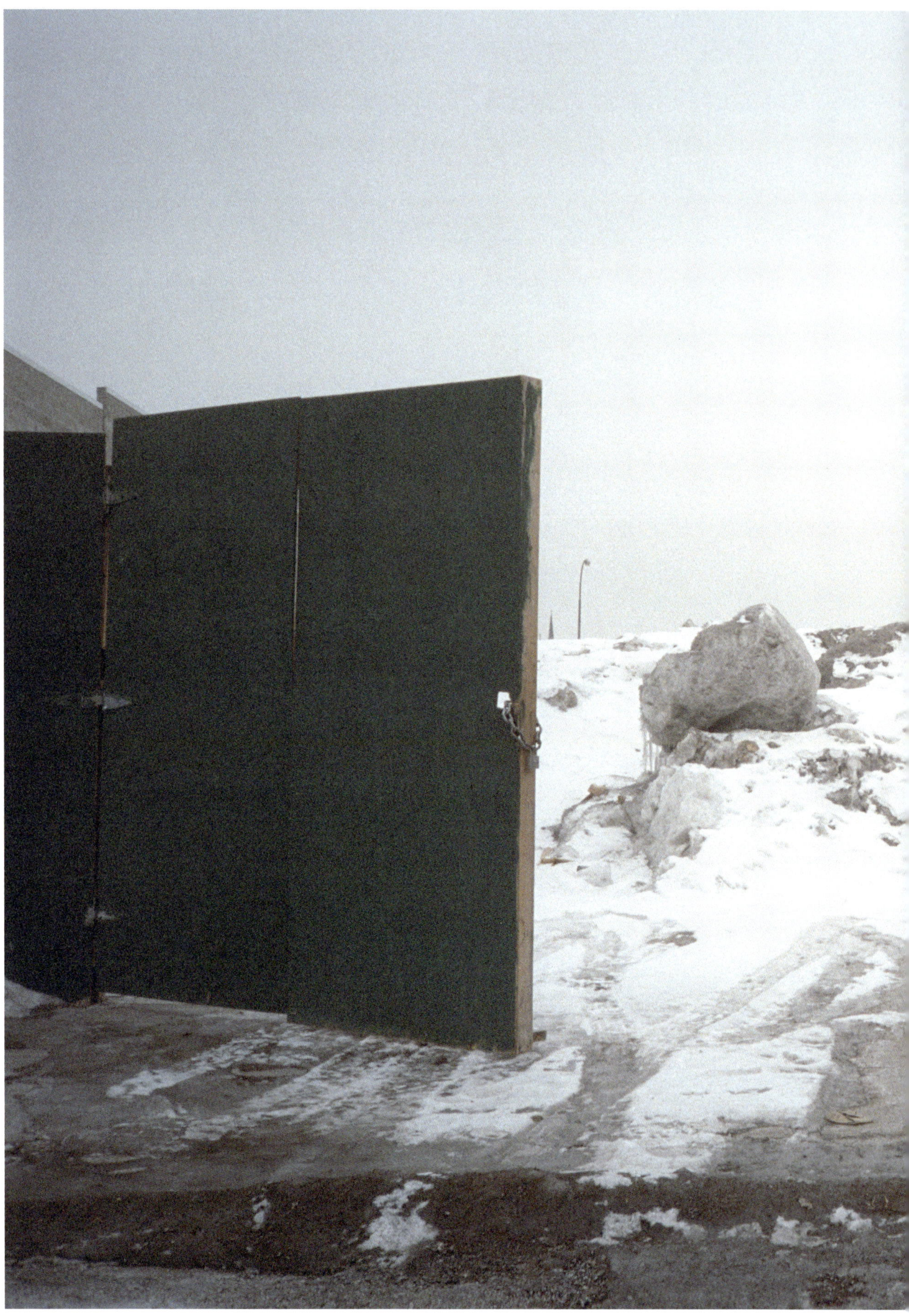

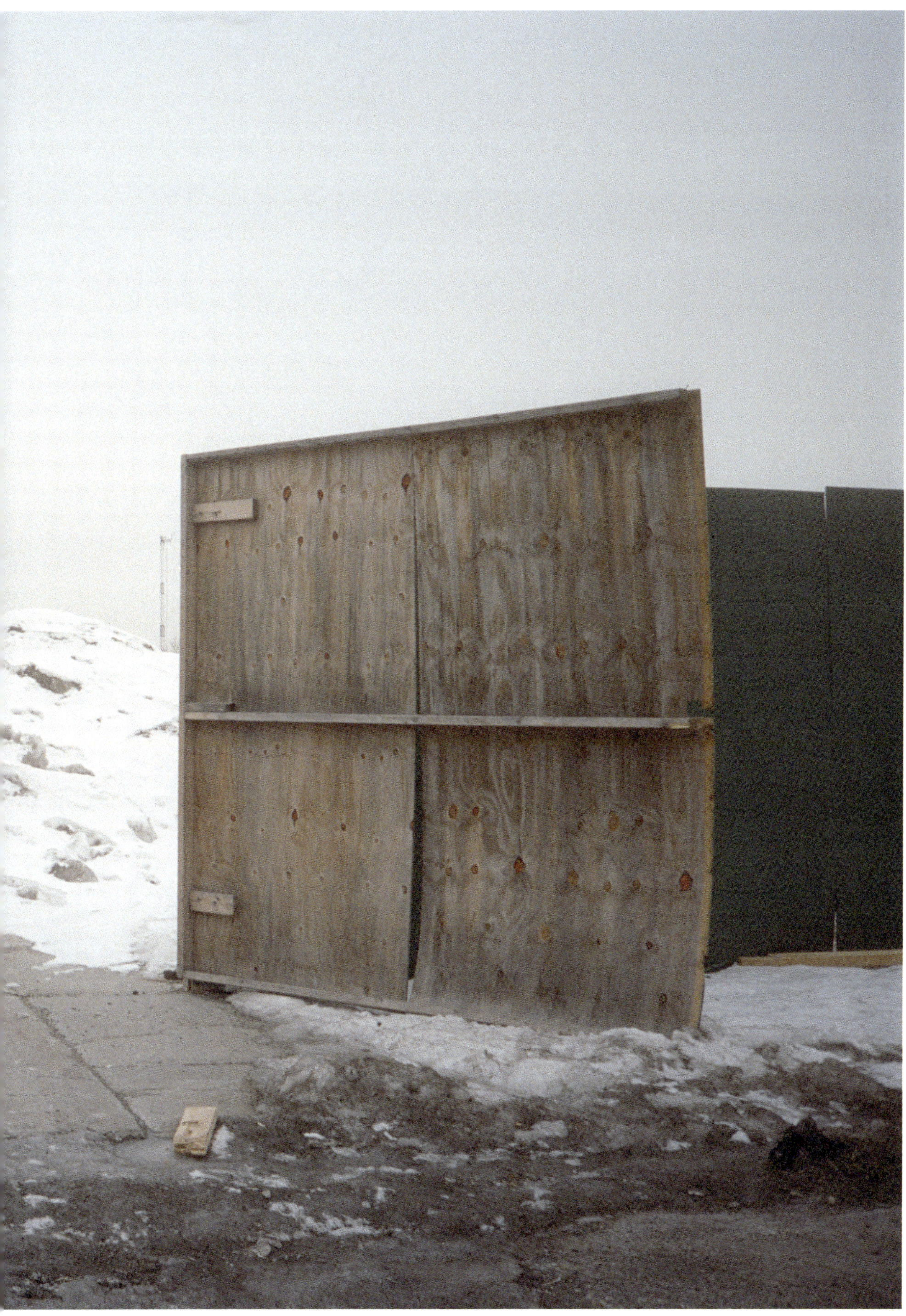

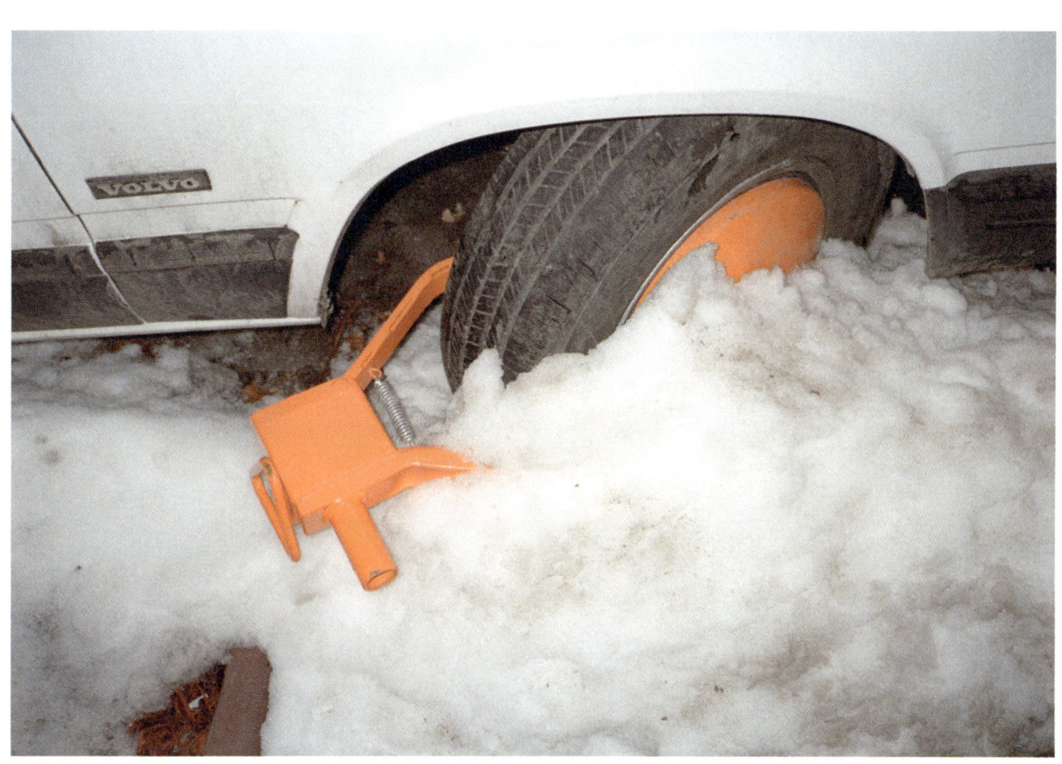

VOLVO

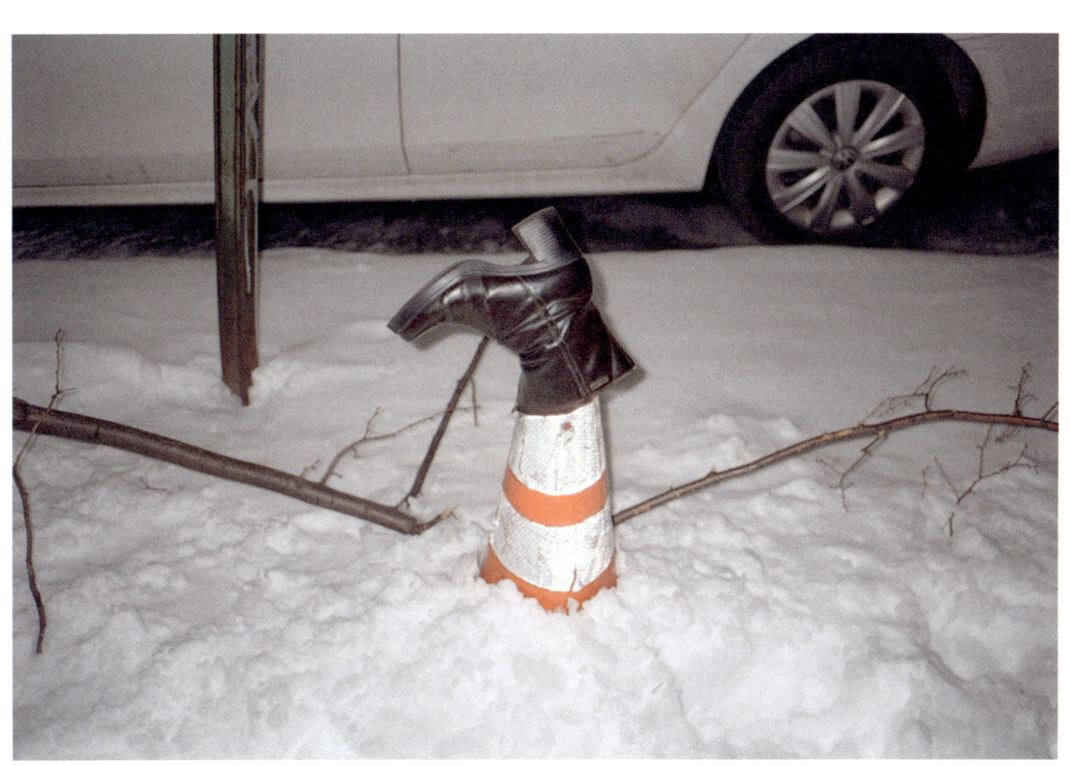

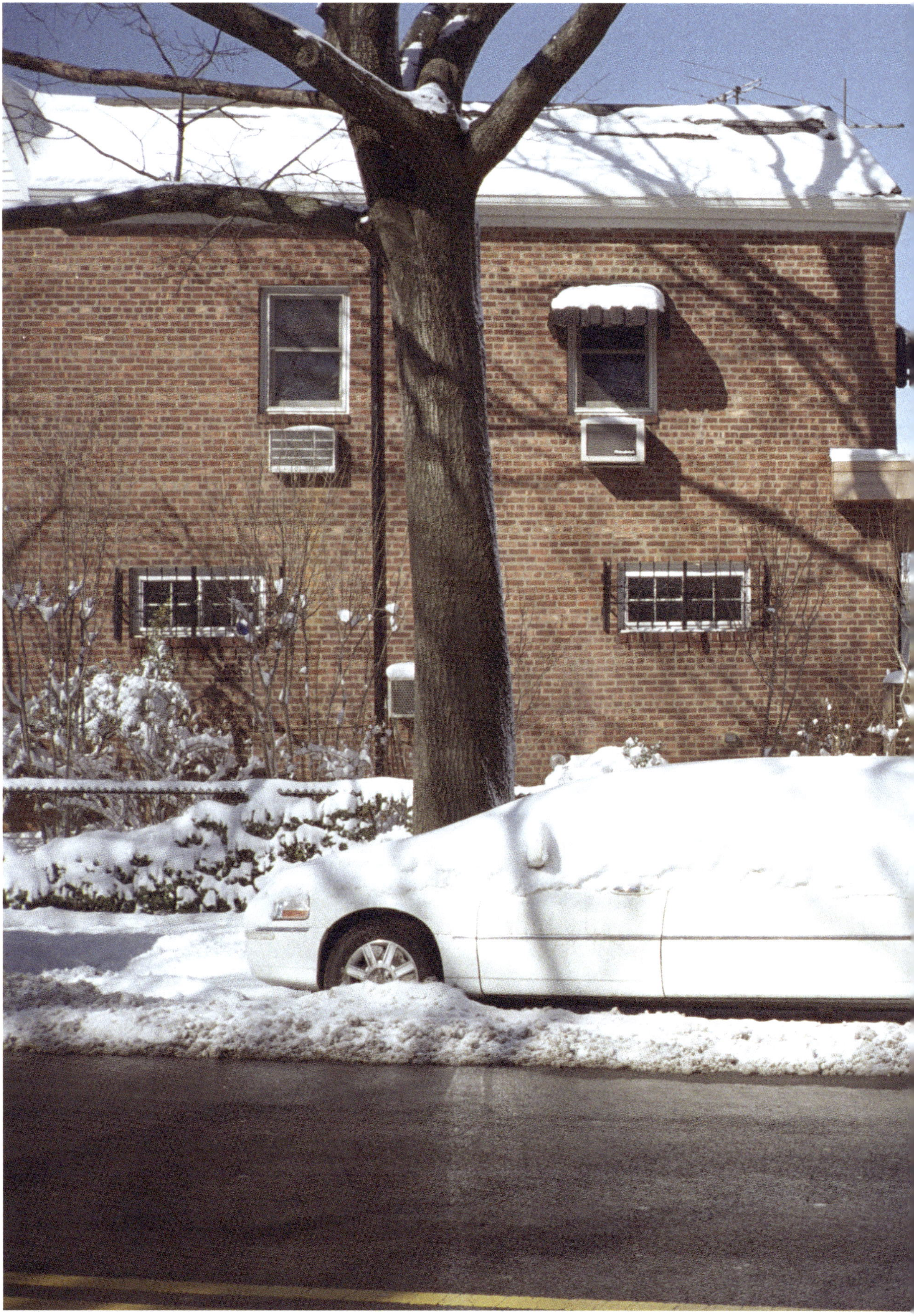

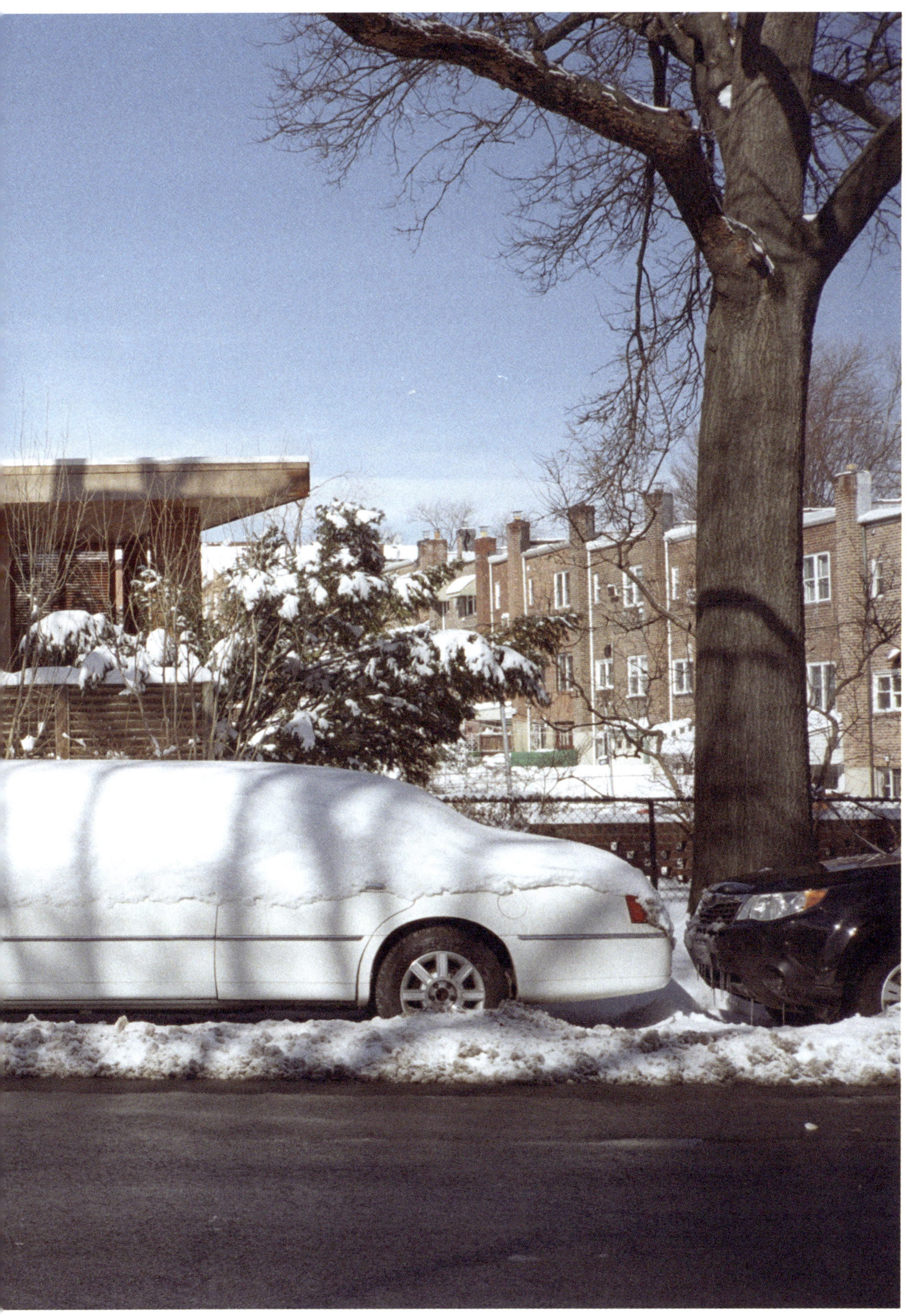

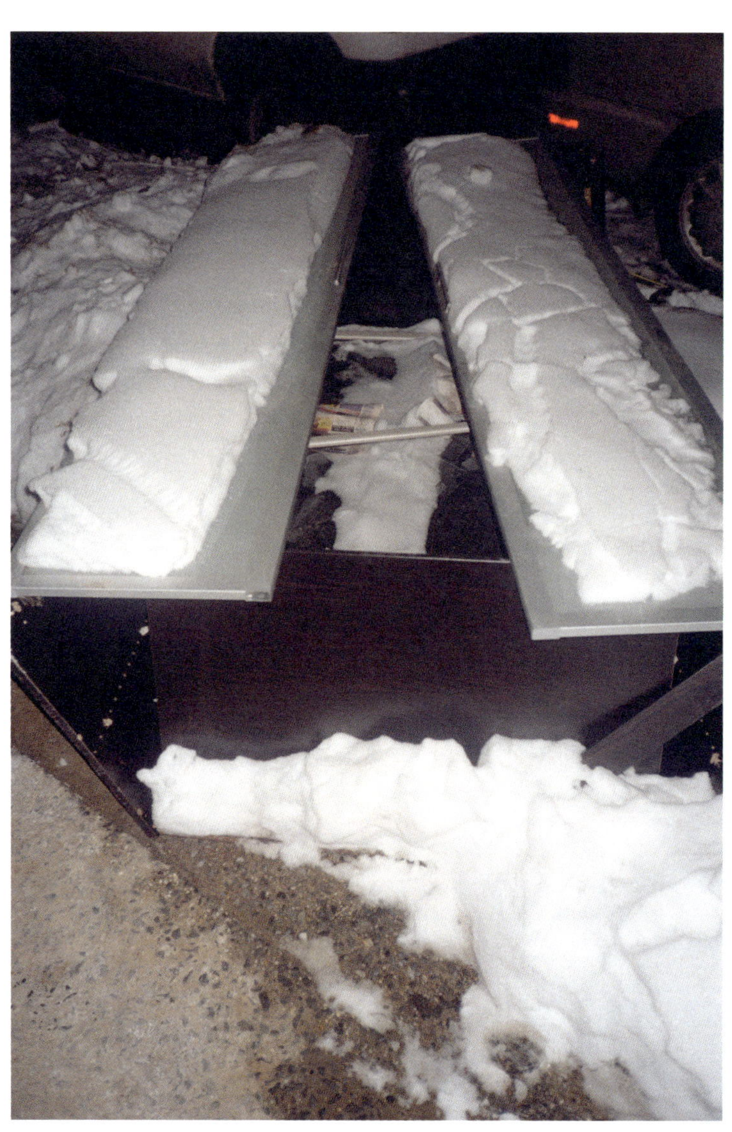

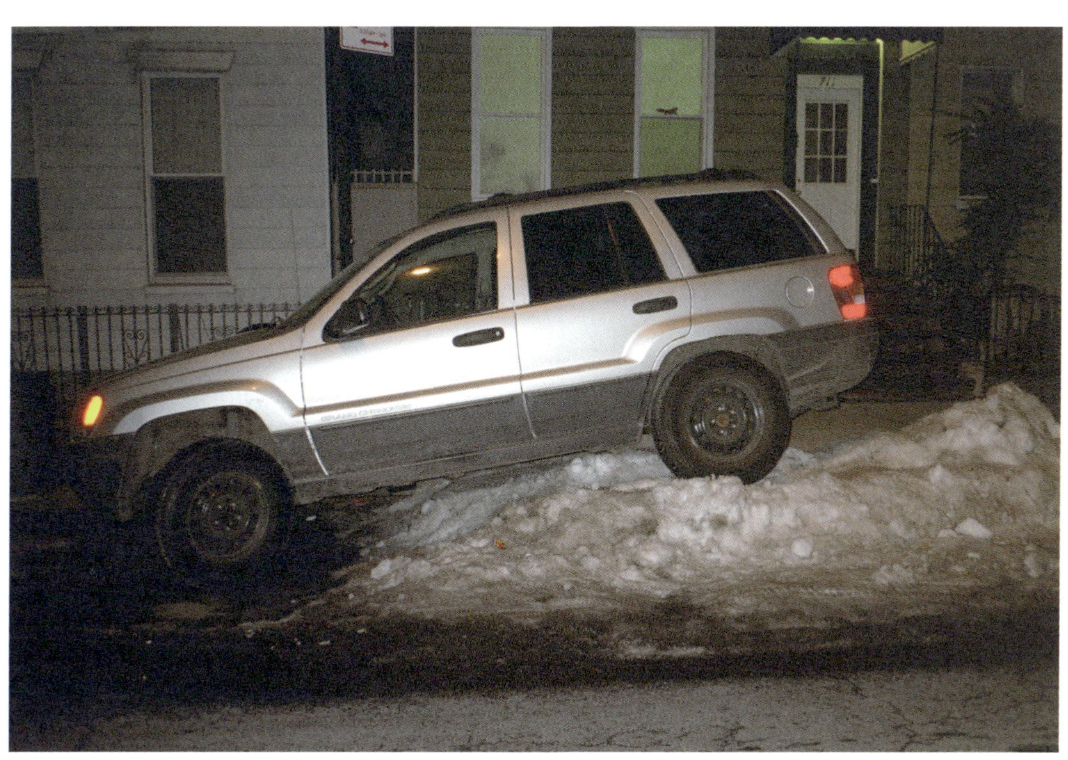

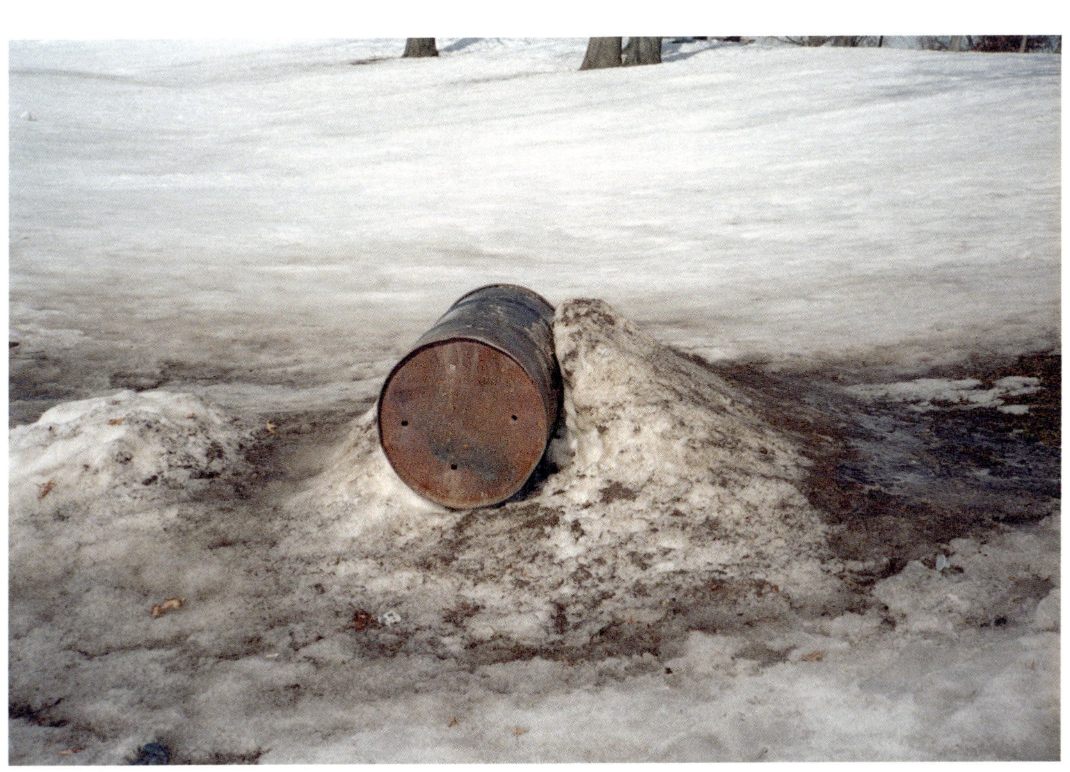

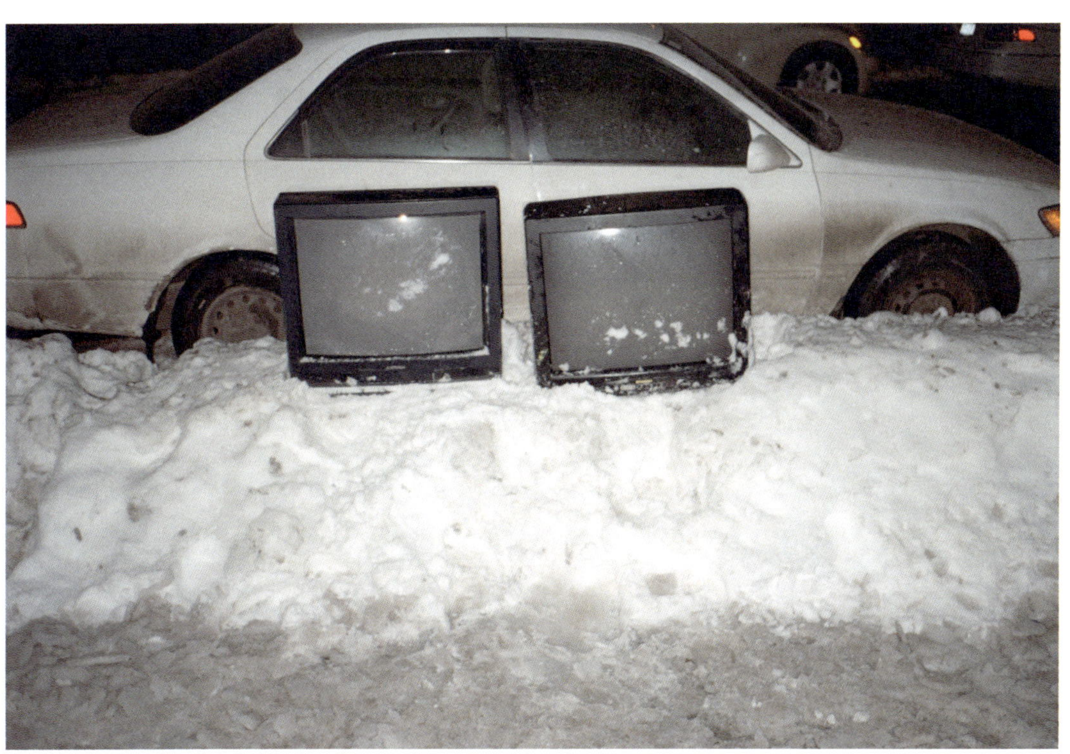

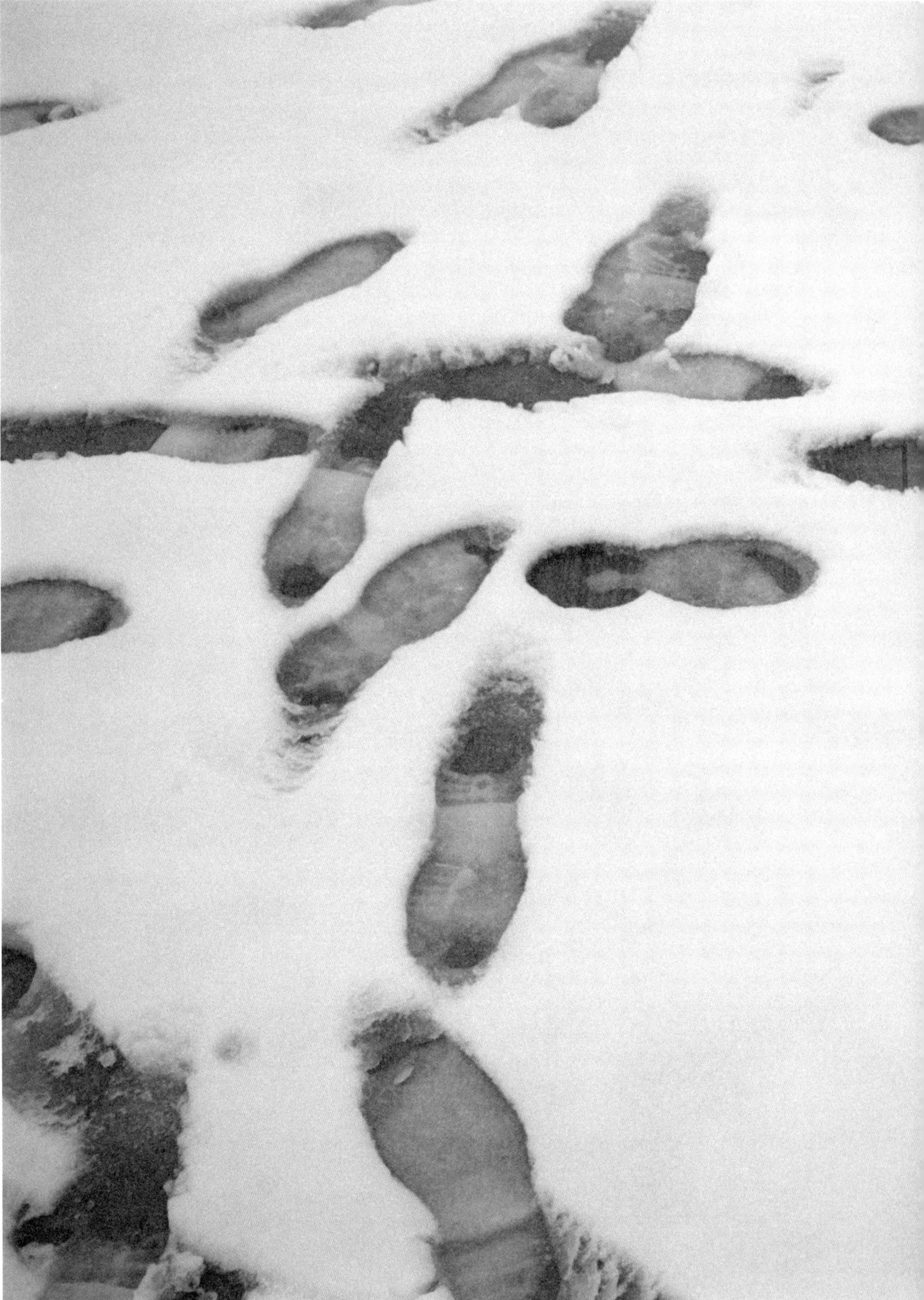

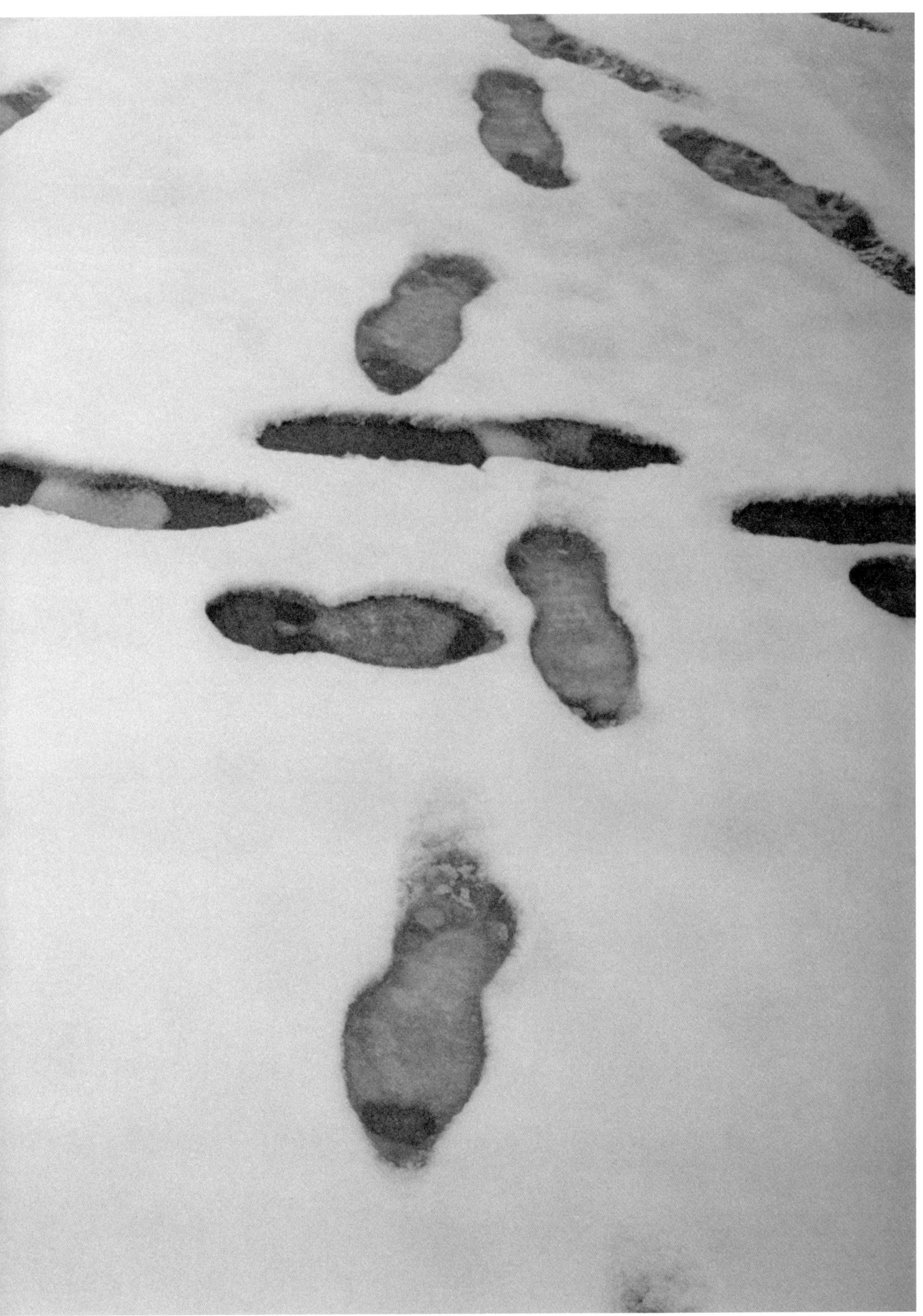

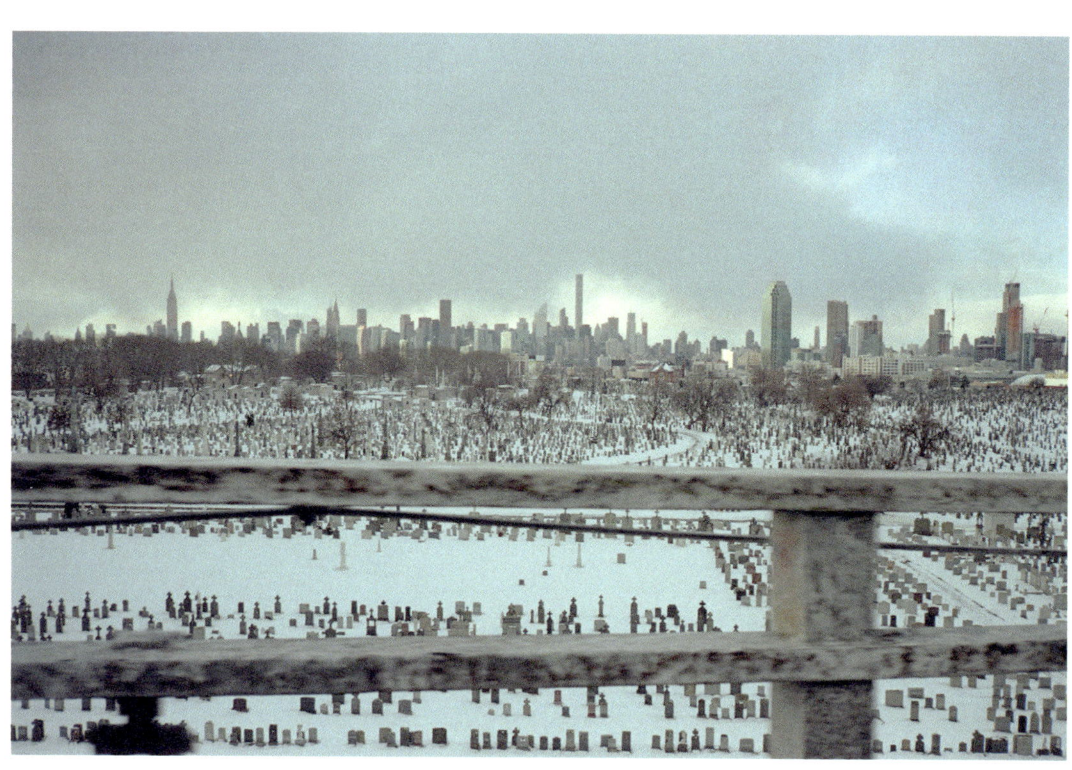